DIAGRAMMATIC KE
PASSENGER SERVIC

Through Services
Other lines
Steamer services

The Western Region before Beeching, based on the timetable map of 1959. The dotted areas were the subject of other, more detailed, maps.

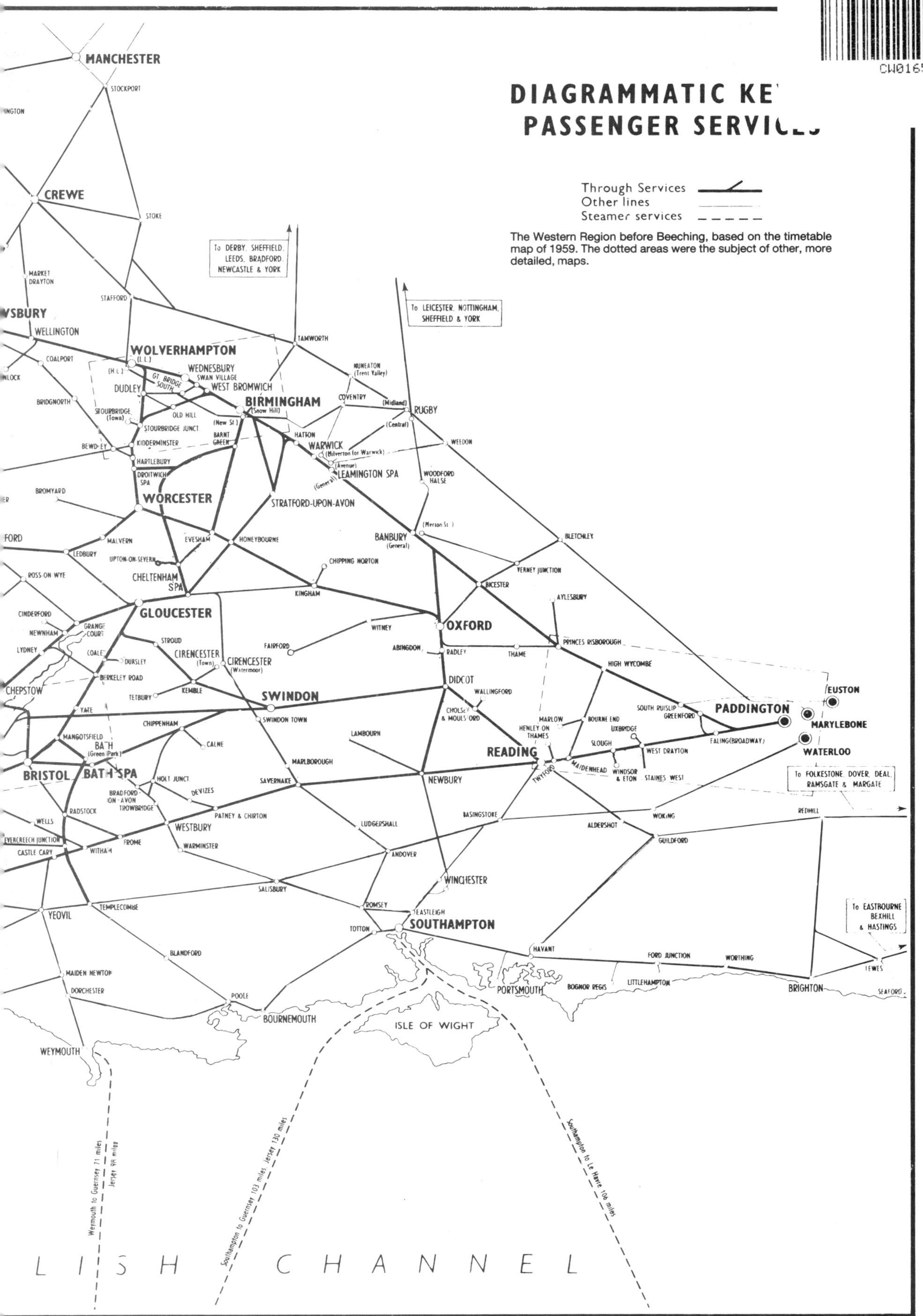

The WESTERN BEFORE BEECHING

NOT IN
COMMON USE

The WESTERN BEFORE BEECHING

Chris Leigh

LONDON

IAN ALLAN LTD

First published 1990

ISBN 0 7110 1898 7

Published by Ian Allan Ltd, Shepperton, Surrey; and printed by Ian Allan Printing Ltd at their works at Coombelands in Runnymede, England

Front cover:
Scenes like this did not survive Beeching. Note only did steam disappear but connecting branch line services ended too, often followed by closure of wayside main line stations and eventually by the demise of mechanical signalling. At Yatton on 18 August 1963, '2251' 0-6-0 No 2268 has arrived light engine from Bristol. A Derby suburban DMU waits in the bay for its next journey down the short branch to Clevedon. Beeching slated this service for closure and it was withdrawn in October 1966.
Hugh Ballantyne

Back cover, top:
Dieselisation alone could not save lightly used rural lines, but even so the aged ex-GWR railcars were kept busy into the early 1960s and were moved around to replace steam workings, finally being concentrated around Worcester. In August 1961, car No W28 which survived long enough to receive green livery, is seen at Cleobury Mortimer on a working between Tenbury Wells and Bewdley, a service which ended a year later, just before the Beeching report.
J. H. Moss/Colour-Rail

Back cover, bottom:
Recently 'soled and heeled' at Newton Abbot works, '57xx' 0-6-0PT No 5796 stops for the guard to pin down the brakes on the loose-coupled wagons of the Kingswear pick-up goods at the top of Torre Bank in April 1957. Beeching brought an end to pick-up freights, loose-coupled wagons, and hastened the end of steam. He also ended steam overhauls at Newton Abbot.
P. W. Gray/Colour-Rail

Contents

Left:
The character and atmosphere of the pre-Beeching railway are summed up in this dismal wet view at Dolgellau on 13 January 1965. The Barmouth-Bala train has paused for a few moments and ahead of it are semaphore signals, water tank, gas lamps, fire devil and two signalboxes, all the trappings of a railway scene soon to pass.
Andrew Muckley

Right.
The large Prairies were the mainstay of passenger services between Cheltenham and Kingham and this view shows No 4141 entering Bourton with a train from Kingham on 28 May 1962. The fine Cotswold stone station building is marred only by the austere 1930s canopy valancing, but otherwise this country station is a glorious mixture of architectural styles. There is a neat and rather pretty waiting shelter, a standard red brick GWR signalbox, and the original black timber goods shed. Note the tall 'gibbet' concrete lamp standards with their wire arrangement for raising and lowering the Tilley oil lamp. When I last visited Bourton in the early 1980s these lamp posts were still there although a road now follows the line of the railway. *Ian Allan Library*

Endpapers:
The Western Region route map before (front) and after (back) the Beeching era.

Introduction

In the summer of 1986 I started building a model railway layout, my first based on British practice for 15 years or more. I still favour modelling real locations rather than inventing some fictitious station to fit a particular layout shape. From time to time I had been tempted to launch into such a project and places such as Chipping Campden, Handborough, Lechlade and Adderbury had seemed likely choices. In the end, I went for somewhere really insignificant, Black Dog halt on the Chippenham-Calne branch line.

Apart from the unusual building, what really appealed about Black Dog was its setting. It would be possible, within a modest layout length, to show the little halt and its single siding within its rural setting and with the trains approaching through open countryside. I wanted to recreate the place in miniature complete with the glimpses through the trees from the main road, which is the way I remembered it.

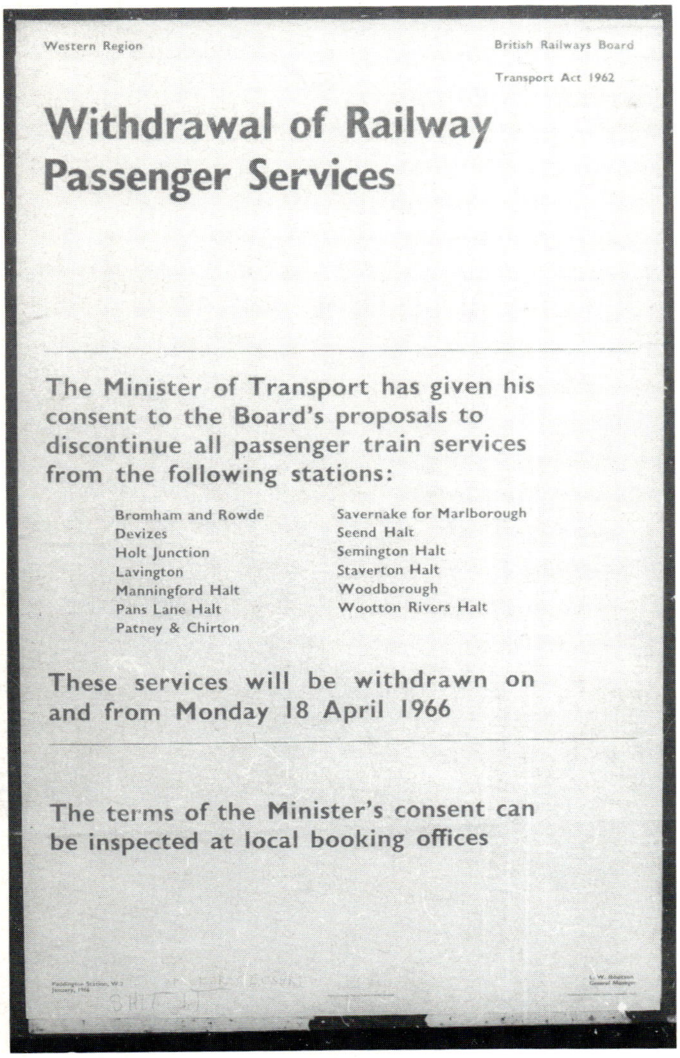

In order to make it as accurate as possible in colourings and atmosphere it was necessary to visit the site to try and recapture the 'feel' of the location. The Calne branch closed in 1965, a victim of declining fortunes whose fate was sealed in a document called *The Reshaping of British Railways* — more usually termed the 'Beeching Report'.

Twenty-plus years had taken their toll. The halt had been demolished, little more than a mound of earth indicating where the platform had been. The iron bridge over the Bath Road had gone and the embankment on one side had been cut back. Nevertheless, the two entrances remained little changed, one with brick walls and standard GWR wooden gates, had been the carriage entrance, the other with wicket gate opening on to a muddy track had been the pedestrian entrance. For Black Dog was one of those Great Western curiosities, a station built for the local landed gentry and not included in public timetables until 1952.

The year 1952 saw the release of *The Titfield Thunderbolt*, the film in which the local squire saves his village branch line from closure. In reality things were usually very different. Black Dog had ceased to be used by the people for whom it was built. They now went by car. Its siding was removed in 1960 because the goods now went in and out by road. Closure of the RAF stations at Yatesbury and Compton Bassett meant that their substantial contribution to passenger traffic on the Calne branch ceased. Increasingly, the branch's other traffic source, the output of C. & T. Harris's Wiltshire bacon and sausage factory in Calne, also went by road. So, in the end, the line closed on a chilly Saturday evening in late September 1965.

I grew up and developed my railway interest during the years immediately prior to Dr Beeching's report. Suddenly, in the midst of all the pleasure of visiting these magical rural railways came 'Beeching'. Almost overnight, it seemed, closure, rationalisation and rundown were everywhere. On limited funds it just was not possible to visit everywhere before closure. I was already trying to cope with pre-Beeching closures such as Much Wenlock and the Midland & South Western Junction line before they were demolished. After Beeching it was not unusual for six or seven different services, spread over the Western Region, to close on the same weekend.

Of Beeching himself, I cannot say much. He seemed to me to be merely the personification of a policy, the poor soul who was the front man for a scheme that few people were going to like. Certainly, he has been vilified by some writers. In fairness he was just a man doing a job the way he felt it needed to be done. Unfortunately he came across as an unsympathetic person who seemed to have little interest in the railway system which he controlled.

Left:
The notice which said it all. Beeching closures in the area west of Bedwyn to Westbury and Devizes. *P. Strong*

Above:
A September 1963 scene at Llanidloes. Fallen lamps and signs are the clues which reveal the truth about the tranquil scene in the background. Demolition has commenced. *Andrew Muckley*

There are those, railwaymen among them, who feel that the good Doctor's medicine, however distasteful, was the only way of saving the railway. Equally, I have talked to dyed-in-the-wool railwaymen who were horrified by what went on in the 1960s and who watched people build their careers on how efficiently and quickly they could drive away traffic and customers and eliminate parts of the rail system.

Beeching is synonymous with closures. In fact, he brought about other changes of considerable significance. The removal of the 'common carrier' requirement and the introduction of freight railheads and merry-go-round block trains also reshaped the Western Region.

It is not the intention of this book to cover in detail the changes wrought by Beeching and his minions. That has been ably done in other publications such as G. Freeman-Allen's *The Western Since 1948*. Nor is it my intention to dwell on photographs showing the 'before' and 'after' effects of the reshaping. In some ways this book is a development of the *GWR Country Stations* formula; indeed the germ of the idea was a third volume in that series.

The change of title and the move to a larger format has enabled a broader look at the Western Region of the late 1950s and early 1960s. As well as stations and railway buildings it can encompass infrastructure and services and look at urban stations and major locations. The theme is essentially one of nostalgia for the railway system before the spate of closures and the other reshapings of the 1960s. There are sections covering in detail specific locations which were particular favourites and which I hope will be of value to modellers.

The availability of two photographic collections has been extremely helpful. Both are names which have featured in my earlier books, yet neither had published more than a tiny fraction of his material before. Special thanks are due to Andrew Muckley for making available his prints with so few strings attached.

Chris Leigh
Old Windsor 1989

It was a time of 'last trains' and brief ceremonies with an air of gloom. On 20 September 1964 '14XX' 0-4-2T No 1444 has brought two auto-trailers from the Gloucester-Chalford service to Swindon to form a Great Western Society railtour. It was bound for the Chippenham-Calne branch where it was destined to be the last steam-worked passenger train. *M. Pope*

CHAPTER 1

The Gathering Storm

The Beeching Plan — or indeed any plan for the reform of the railways — can be viewed as an inevitable result of the amalgamation of the 'Big Four' companies. In the years following Nationalisation there was a primary need to bring the constituent 'regions' together as a coherent railway operation. From this there followed standardisation of equipment and procedures. There then came the urgent need for modernisation following the years of wartime neglect. Inevitably, anyone looking at such modernisation in the 1950s would turn towards diesel or electric traction, rather than steam. So, the year 1955 and the launch of the Modernisation Plan, which heralded the end of steam, is an important landmark in BR history.

The benefits of modernisation were needed quickly in order to combat ever-increasing requirements for financial support from the Government, and so the Modernisation Plan envisaged a fairly quick changeover, with steam eliminated by the mid-1960s. Modernisation alone could not produce the desired economies even though it would bring reductions in staff levels. The railway system had come to be visualised as a giant tree, full of dead wood which merely soaked up vast amounts of public money. Thus, when Dr Beeching launched his reshaping plan he used this 'tree-surgery' analogy, particularly with regard to the withdrawal of loss-making passenger services.

Of course, closures of stations and withdrawal of unprofitable services were as old as the railways themselves. In the West Midlands, for instance, passenger services over

Below:
On the branch lines, working practices remained as they had been for many years, though in places the abolition of station staff, gatekeepers and signalmen made extra duties for the train crew. Here, the guard of the Tiverton-Hemyock branch passenger train walks forward to open the level crossing gates at Culmstock for the passage of his train.
Ian Allan Library

the branch to Halesowen had been withdrawn in the 1920s. Despite being in a well populated area the service had failed to attract sufficient patronage — largely because of effective competition from local buses. In the West Country, the attractive and very scenic Lynton & Barnstaple line had been closed in 1935. Here, although the railway had enjoyed healthy summer tourist traffic, receipts were insufficient to carry it over the lean winter months. When track renewals became necessary, they could not be justified financially and so the line closed. Road competition and the fact that the line was to narrow rather than standard gauge, were lesser factors in the decision.

By the 1950s both the factors of road competition (from the private car even more than the bus) and the need for expensive renewals following years of neglected maintenance, had reached epidemic proportions. The problems were worst in rural areas and where lightly-used local services were slow or infrequent. So it was that the pace of closures and passenger service withdrawals increased through the 1950s. On the Western Region (WR), where there are large tracts of rural territory, many closures took place during this period, before the Beeching Report.

Victorian enthusiasm for railways had left the WR a legacy of rural lines with very limited traffic potential and with wayside stations which did not serve significant local population. Pre-Beeching branch line closures in rural areas included:

Malmesbury-Little Somerford	September 1951
Leominster-Bromyard	September 1952
Monmouth — Ross-on-Wye	January 1959
Moretonhampstead-Newton Abbot	March 1959
Craven Arms-Much Wenlock	December 1959
Lambourn-Newbury	January 1960
Cardigan-Whitland	September 1962
Fairford-Oxford	June 1962

Right and below:
Express-rated goods vehicles could be added to certain passenger workings, a practice which continued into the DMU era. Here, an Exe Valley line push-pull working to Dulverton waits at Exeter St Davids with '14XX' No 1450 propelling auto-trailers Nos W225/8 and with a wooden-bodied horsebox attached to the rear. *Ian Allan Library*

These are just a random selection spread over the 10 years leading up to the Beeching Report, and there were a number of others. Two routes which cut through the WR, from north to south, the ex-Didcot, Newbury & Southampton line and the former Midland & South Western Junction Railway from Cheltenham to Andover, also closed just before Beeching. Among the others mentioned above, some, such as Craven Arms-Much Wenlock and Leominster-Bromyard, were sections of through routes which left truncated branch lines to soldier on for a while longer before they too, succumbed to the inevitable.

During this period there were a few attempts to promote patronage, or at least to take advantage of passenger potential which was thought to exist. On branch lines this generally meant the erection of unstaffed halts, as at Poyle Estate and Colnbrook Estate on the Staines West branch. These two were built to serve industrial estate developments. In practice they resulted in good loadings on one morning and one evening train but it is doubtful whether the resulting fares ever covered even the construction costs of the concrete platforms.

Dieselisation was comparatively late on the WR and accordingly any benefits which it brought to branch lines and local services were a case of 'too little, too late'. Elsewhere on BR the influx of 'Derby Lightweight' diesel units began in 1954/5, but the WR had only its handful of ageing GWR railcars until deliveries of suburban units began in 1957/8. It was not until major deliveries of Pressed Steel units in 1960 that DMUs could be cascaded in any quantity to branch line and local services and by then the fate of many rural lines was sealed. Single-unit railcars did take over some lines such as Didcot-Newbury, before closure, but others such as the Fairford branch remained steam-hauled to the end. However, such was the apparent lack of communication over the future of some routes, that destination blinds for diesel railcars included names of closed lines. Years after Fairford closed it was possible to see DMUs displaying this destination, despite the fact that they had never been there!

Among diesel passenger vehicles the cheapest option was the four-wheeled diesel railbus. The idea was based on successful German practice. Four vehicles were allocated to the Western Region and operated 'experimentally' on the branches from Kemble-Cirencester and Tetbury. On both branches new halt platforms of very basic nature were built to serve villages, housing estates, and — in one case — a public house. The old closed station at Culkerton was reopened as an unstaffed halt.

The whole thing must have been doomed from the start. The branches differed considerably in character, Tetbury being a quiet village offering little traffic, while Cirencester was busy and offered much more potential. To link the two operationally and commercially would inevitably mean that one would pull the other down. It was quickly found that the lightweight railbuses could not be relied upon to operate track circuits. They therefore could only be used empty when operating over main lines and thus ended any possibility of operating through services from the branches to Swindon. The operating diagrams employed two units in traffic, one on each branch. A third car ran empty from Swindon in mid-morning to relieve one of them, which then returned to Swindon for fuelling etc. The fourth car was kept as spare and for cover during overhauls etc. Four vehicles to cover a two-vehicle service scarcely constituted the kind of intensive use which might have made for economical operation.

The branches from Kemble were both targeted by Beeching and closed within a year of his Report being published. Thus came the Beeching concept of 'railheads' — points to which people would travel to join a main line train. Kemble was to remain as a railhead for the area. The adjacent village comprised only a handful of houses, so there would be negligible local traffic. The railhead strategy envisaged that passengers would come by car to Kemble from Cirencester and the surrounding area, to join a fast service from Kemble. This service would be speeded up, since a number of stops at local stations would be eliminated. In practice, once in his car, the passenger would travel to Swindon before joining his train — or he might even use his car for the whole journey. This was the flaw in the 'railhead' scheme. The follow-on from the decline in use of 'railhead' stations was a reduction in services over the main line. Taking Kemble as an example, the whole route from Swindon-Gloucester has become little more than a branch line, single-tracked for much of its length, with only two staffed intermediate stations (Kemble and Stroud), and with much of its service comprising of a DMU shuttle.

There were, too, anachronistic operating practices which the WR inherited. At Farington Gurney halt, tickets were sold at the village pub, a bell being provided to alert the landlord outside licensing hours. At Appleford the village post office provided the service, while at Dilton Marsh a notice directed passengers to a private house to buy their tickets, an arrangement which lasted long after the Beeching Report.

What the Beeching Report did, as far as passenger services were concerned, was to set out to eliminate all those services which were not viable and to close lines and individual stations in order to reduce the deficit. Savings would result from reducing staff numbers accordingly and land and buildings would be released for other uses.

A look at the figures of track mileage on the whole of BR (individual figures for the WR were not available) shows just how the pace of closures accelerated. In 1948 BR took over 19,598 route miles of railway. In its first 10 years just 1,000 route miles were closed, a goodly proportion of them on the WR. In the next five years up to 1962, a further 1,400 miles were closed. In 1963 alone, 500 route miles went, and in 1964, the first full year after the Beeching Report, another 1,000 miles were slashed. Closures under Beeching continued through 1965/6 until the bulk of them had been dealt with, and the pace then slackened off. Service withdrawals, at least on the WR did continue into the early 1970s, with controversial losses such as Barnstaple-Ilfracombe, and Bourne End-High Wycombe which severed another useful through route. The latter was not originally envisaged by Beeching, as neither was Exeter-Okehampton which was the result of traffic decline following the loss of its through route status. Clearly, the truncated Bourne End-Maidenhead section could have gone the same way, were it not for solid local efforts in the late 1970s.

Thus we can see that the 'medicine' of the Beeching closures did often produce side effects which brought other services into jeopardy. Although not part of pre-Beeching Western territory, all the ex-Southern Region (SR) lines in the West of England came into the Western Region concurrent with the Beeching Report. In North Devon and Cornwall, all ex-SR routes beyond Barnstaple were eventually closed, leaving the Exeter-Barnstaple route itself as a straggling branch line which, in the 1980s, is scarcely able to sustain half a dozen trains a day.

Above and left:
A bird on the up platform fence watches an Oxford-bound Pressed Steel DMU glide into Appleford halt near Didcot on 11 April 1967. The halt is still pure GWR, with timber-framed sleeper platforms, Tilley oil lamps and GWR 'pagoda' shelters. Tickets could be purchased from the village post office. Appleford halt was not proposed for closure by Beeching and despite some Network SouthEast modernisation, remains substantially unaltered to this day. *Both: Andrew Muckley*

Below left and above right:
Even more unusual were the arrangements for ticket purchase at Dilton Marsh halt near Warminster. With a platform similar to that at Appleford, but a simple wooden shelter and Southern Region nameboards following its transfer to SR control, it too has survived Beeching. With so many stations reduced to unstaffed status and stripped of their buildings following Beeching, the words 'halt' and 'platform' were dropped from the railway vocabulary. *Both: Andrew Muckley*

The Reshaping of British Railways was a report by the British Railways Board, under Chairman, Dr Richard Beeching, to the then Minister of Transport Mr Ernest Marples. The fact that the government of the day was a Conservative one, and that the Minister of Transport had strong links with the road haulage industry (he was of the Marples family whose firm Marples Ridgeway built the elevated section of the M4 motorway into London) inevitably led one to the conclusion that the Report would have been slanted to say what the Government wanted to hear. In fairness, when the government changed and Mrs Barbara Castle became Minister of Transport, there were still plenty of 'Beeching' closures in progress which could have been refused ministerial consent. In practice, very few were refused.

The Report listed various problems facing the railways, looked at ways of overcoming them and proposed a series of steps to achieving this end. It was said that, if implemented, they would almost eliminate the railway's deficit by 1970. The reasons why, 20 years on, the railway was still working to eliminate its deficit are outside the remit of this book. With the numbers of station reopenings and the upswing in traffic on BR, plus the perennial moans about the amount of heavy goods traffic forced on to our roads, one could certainly make a pretty convincing argument on the theme that Beeching's medicine did not work.

The 15 steps proposed in the Report were:

1 Discontinuance of many stopping passenger services.
2 Transfer of modern DMU stock displaced by (1) to replace steam working on continuing services.

Below and left:
Another unusual service, which ended with the opening of the Severn road bridge, was the Pilning-Severn Tunnel Junction car train. Usually formed of a single coach and just a couple of wagons, it ran through the Severn Tunnel enabling motorists to avoid the traffic congestion in Gloucester or the long wait for the ferry. The photographs show loaded car flats in the bay siding at Pilning, and the 4.40pm working to Pilning climbing away from the tunnel on 9 May 1958 with '61XX' 2-6-2T No 6119 in charge. *M. Mensing (inset) M. Leigh*

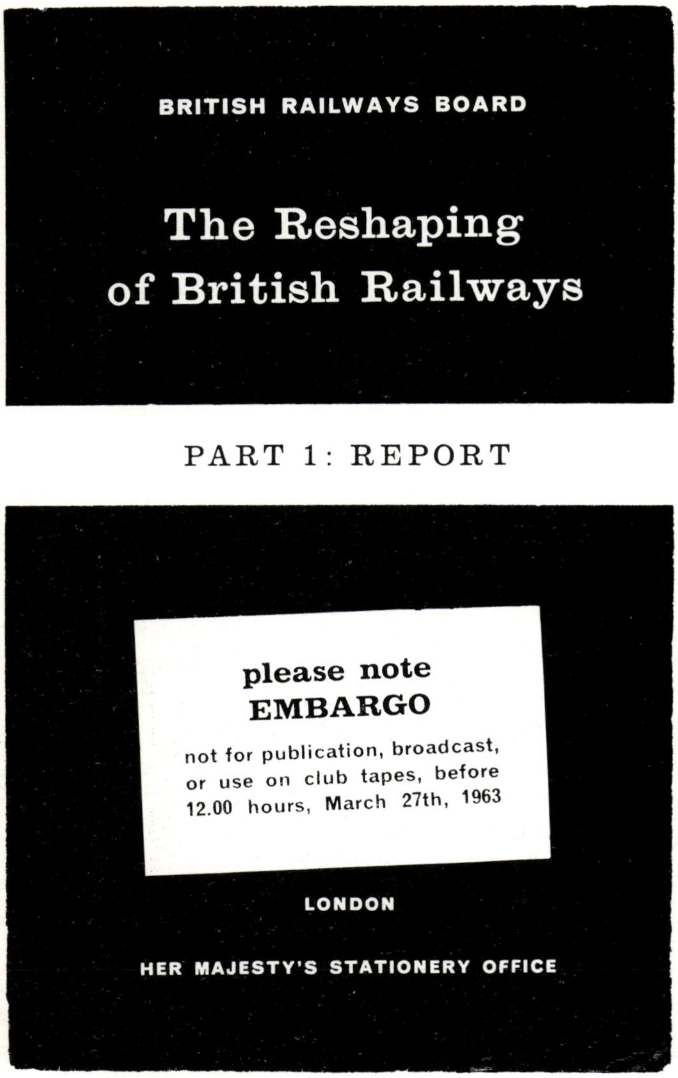

Above:
The jacket of the Beeching Report

3 Closure of a large number of small stations to passenger traffic.
4 Damping down of seasonal passenger traffic peaks, enabling the withdrawal of corridor coaching stock held to cover them.
5 Improvement of Inter-City passenger services and rationalisation of routes.
6 Co-ordination of suburban train and bus services and charges.
7 Co-ordination of passenger parcels service with the Post Office.
8 Increase block train movement of coal by inducing the NCB to provide loading facilities and by providing coal concentration depots for domestic distribution.
9 The reduction of uneconomic freight mileage by closures of small goods stations, and consequent adjustment of charges.
10 The ending of the wagon forwarding system and its replacement by siding-to-siding traffic operated by through trains.
11 Study and development of the 'liner-train' (container) system.

Right:
Beeching valued the parcels and mail business but wanted to streamline arrangements with the Post Office in order to reduce costs. Coincidentally, improvements in motive power and the pathing of trains at night meant that mailbag collection on the move by TPO trains could be replaced by routine stops. So ended the colourful practice of picking up and dropping mail at lineside locations marked by a chequerboard as here at Maidenhead. 'Hall' No 5967 *Bickmarsh Hall* hurries past with an up express. *D. B. Hart*

12 Concentration of freight sundries traffic on 100 main depots.
13 Massive withdrawals of freight wagons over the three years following the Report.
14 Continued replacement of steam by diesel locomotives on main line services, requiring in total some 3,750-4,250 locomotives.
15 Rationalisation of the railway's road cartage fleet and operations.

With benefit of hindsight it is easy to see the flaws in the Report and to understand why it is seen today in a less-than-favourable light. Although its aim was to enable the railway system to do profitably the things to which it was best suited, it gave the impression that there was little to offer beyond closures and contractions.

In the early 1960s, public relations was not considered an important part of business practice in the way that it is today. There is no doubt that the Report, publicised as it was as a huge list of closures, was disastrous public relations for BR. There was scarcely a community of any size unaffected by its provisions. Its positive points such as the furtherance of block train operations and its involvement in the novel international containerisation system had little or no significance to the man in the street. One of the principal factors in the establishment of 'corporate identity' and the new-look BR of the late 1960s, was the urgent need to overcome the public relations damage done to the railway's image by the Beeching era.

The following chapters will look more closely at some of the different aspects of the WR which changed either as a direct result of the Beeching Report, or took place simultaneously with its implementation. To end this chapter I can do no better than to quote from the *Railway Gazette* of 29 March 1963, just two days after publication of the Report, for this was the beginning of the end of the Western Region I knew and loved. Headed 'Political dynamite!', the *Gazette's* editorial read '. . . The proposals themselves are sufficiently drastic to call for either great courage or a spirit of desperation on the part of the Government, which will have to decide, whether to implement them in whole or in part. It would seem clear that the plan is not of a kind which would lend itself to much in the way of dilution or compromise if the objective of making what remains of the railways a payable unit is to be achieved.'

Right:
Another view of the lineside TPO apparatus showing the traductor arm which would be loaded with mailbags and turned outwards for collection, and the net for receiving bags dropped from the train.
D. B. Hart

5967

CHAPTER 2

Along the Main Lines

In the list of 15 points from the Beeching Report, the first item, the discontinuance of many stopping passenger services, was inextricably linked to item 3, the closure of small passenger stations. Closures of stations brought reductions of staff with consequent savings, and also enabled sites to be sold off. The withdrawal of services from such stations enabled old rolling stock to be scrapped or new DMUs to be cascaded, in modern parlance, to other services, in turn replacing stock which could be scrapped.

The criterion for withdrawal of a service was its inability to produce sufficient revenue even to cover operating costs. It will not be until after 1993 when we are able to study the financial cases for individual closures as these are covered by the '30-year rule' for release of records. However, the financial cases were based on the then-current accounting procedures and on figures covering receipts in recent years. There was to be no 'last-ditch' initiative to increase revenues on services which were under threat. The statutory closure procedure required BR to post notices of closure which invited written objections. As soon as one objection was received, the posters would be pasted over with a slip stating that closure had been deferred pending consideration of objections. There would then follow a formal notice stating that the Minister had given his consent to withdrawal of the service and listing the stations to be closed. In many cases a year or more would elapse between posting of the first notice and the actual closure.

Below:
A photographer with his old plate camera and tripod catches 'Castle' No 7007 *Great Western* racing down through Slough with an express in the early 1950s. No 7007 was the last locomotive built by the Great Western Railway and in the decade before Beeching little changed on the Western Region. The majority of trains were still formed of GWR stock hauled by Swindon-built motive power. *Ian Allan Library*

On the main lines, the placing of wayside stations had been a matter of policy which dated back to the earliest days of the GWR. Obviously, important locations along the route would be provided with a station. In between those locations, it was considered desirable to have a local station every few miles, even if there was no significant source of potential revenue nearby. The reasoning behind this was an original 'railhead' concept — namely that rural communities faced with slow road or canal transport as their only alternative, would make use of their nearest local station. Thus, early railway maps included stations with the word 'Road' in their title, to indicate their location on the road to a particular town or village. Some, such as Wantage Road, lasted into the 1960s. In the early days, there is little doubt that these railheads worked. When the horse and bicycle were replaced as local transport by the lorry, tractor, and car, such stations were quick to see their traffic, in freight and passengers, drift away. They should have been the signpost to the fact that railheads had seen their day and trying to recreate them, but fewer and further between, would not work.

The operation of passenger trains to serve these wayside stations involved a network of secondary, or local services, which by the late 1950s were still largely steam-worked using elderly, downgraded ex-GWR coaching stock. Diesel multiple-units took over some such services from the turn of the decade, while on others, Type 2 or 3 main line diesels became a substitute for steam. Two coaches behind a 'Hymek' could scarcely have constituted an economy in operating costs!

Some very lightly-used wayside stations had been closed prior to Beeching, but they were often odd examples which merely eliminated one stop from a service which was making perhaps a couple of dozen calls. What the Beeching Report set out to do was to rationalise such services by eliminating all but the really worthwhile stops. These local services, stopping at all stations, were necessarily slow and, on double-track main lines, they could get in the way of non-stopping expresses. At best, they made timetabling more difficult.

Taking the Oxford-Worcester line as an example, the plan was to close the majority of the wayside stations and halts, leaving railheads at Kingham, Moreton-in-Marsh, Evesham and Pershore. As with all passenger service withdrawals, attention would be drawn to available alternative bus services. If none existed, it would be provided. The strategy for this line did not work, however. The Minister refused consent to the closure of Combe halt, Handborough, Charlbury, Finstock halt, Ascott-under-Wychwood and Shipton, because a suitable replacement bus service could not be provided owing to the local roads system tending to run across rather than parallel with the railway.

With the exception of Charlbury, all were reduced to unstaffed halts and their buildings demolished. Beyond Kingham, consent was given to closure of Adlestrop,

Left:
The Kingswear branch was treated as a main line until the time of Beeching and expresses carried their Paddington-Torquay-Paignton-Dartmouth roofboards well into the 1960s. Dartmouth was reached by ferry from the rail terminus at Kingswear and trains were worked throughout from London by the heaviest express locomotives. Here, 'King' No 6012 *King Edward VI* is turned at Kingswear on 2 October 1958 in readiness for an up working. With Beeching, the line from Paignton to Kingswear was progressively reduced in status, eventually being sold off to the Dart Valley Railway. *Ian Allan Library*

Left:
The busy scene at Kingswear, looking in the opposite direction, on the same date. 'Castle' No 7001 *Sir James Milne* stands at the head of the up 'Torbay Express' formed of Mk 1 coaching stock in brown and cream livery. Locomotive and coaches are smart and well polished as befits one of the Western's principal expresses. In the background the quay is host to a collier unloading coal for movement a few miles up the line to Torquay gasworks.
Ian Allan Library

Blockley, Chipping Campden, Honeybourne, Littleton & Badsey, Fladbury, and Stoulton, while Pershore also lost its staff. The reprieved stations have, to this day, suffered the most minimal basic service of a couple of trains each way per day, purely to cater for commuters into Oxford.

Perhaps the strangest feature of this particular story, however, relates to Honeybourne. A substantial four-platform station, it was the largest station on this section to be closed. Its size had, however, related to its operational importance as a junction, rather than to its traffic potential, there being no major centre of population nearby. Post-Beeching, local development has been such that Honeybourne was reopened in response to public demand in the 1980s, one of the first Beeching closures to be rescinded.

Another aspect of the Beeching strategy on main lines was the recovery of under-used assets. Appendix 2 lists the euphemistically-titled 'Services to be modified'. Didcot-Swindon, for instance, was a four-track main line, the local stations being served only by platforms on the relief (or 'slow') lines. Close the local stations, and you could reduce to two tracks and recover miles of redundant track. On other routes, local stations were served by loops off the main running lines so local trains waiting at the platform could be overtaken by through traffic. Close the stations and the loops could go. Alternatively, if the station were to remain in operation, recast the service so that overtaking manoeuvres were eliminated, and the through tracks could be slewed to serve the platform and the loops removed. Stations such as Chipping Sodbury on the South Wales direct line suffered the former fate, while places such as Gerrards Cross on the Birmingham line received the latter treatment.

Above:
Kingswear had full locomotive facilities for dealing with the largest main line classes. In this view, circa 1960, 'Castle' No 5031 *Totnes Castle* has been turned and prepared for working an up express. In the background 'Warship' No D808 *Centaur* stands at the head of the next departure. *D. B. Hart*

Above right:
Some expresses had become quite light by the early 1960s and on the Paddington-Worcester line few trains warranted more than about seven coaches. In this 1962 scene 'Castle' No 7007 *Great Western* is seen again, this time racing down from Campden Tunnel and passing Chipping Campden with an express for Oxford and Paddington. The locomotive is still smart, but most of the BR Mk 1 coaches are now in the maroon livery. The level crossing gates are in original form and controlled from a handsome GWR signalbox. Chipping Campden crossing is nowadays half-barriers, and of the station hardly a trace remains. It was closed by Beeching in January 1966. *Ian Allan Library*

Right:
Further down the Hereford line was Stoke Edith which had closed to passengers a few months earlier than Chipping Campden, in April 1965. Arriving behind a 'Hymek' Type 3 diesel is a typical three-coach local service which will be added to the Paddington train at Worcester. Such joining and splitting of trains was also against the principles of Beeching. The station building seen here is a good example of a standard GWR design dating from the early years of the present century. Although the brick-built construction, steel-framed canopy and flat roof were good quality, cost-effective buildings in their day, they lacked such features as a damp-proof course and in later years suffered rising damp and dry rot, while large areas of timber were expensive in maintenance. Those which survived Beeching usually lost their canopies in order to reduce maintenance costs. *Andrew Muckley*

BEWARE
OF THE
TRAINS

STOKE EDITH
1A95

Above:
As on the other regions, BR Standard motive power was in evidence on the Western. A handful of 'Britannia' Pacifics were drafted in during the mid-1950s for use mainly on the London-South Wales expresses and also within Cornwall. Here, on a bitter cold night in January 1958 No 70029 *Shooting Star* has paused at Badminton with the down 'Red Dragon' Paddington-Swansea express. Badminton was one of a number of stations over which the local landed gentry had a degree of influence. In this case his Lordship was empowered to have certain trains stopped there if he desired. Badminton closed in June 1968 as part of Beeching's reorganisation of local services. *G. Heiron*

Mention of the Birmingham line raises another aspect of the strategy, namely that of eliminating so-called duplicate routes. These were seen mainly as having been a product of Victorian era competition between railway companies. Thus, the WR had a route from London to Birmingham (two if you distinguish between the direct line and the route via Oxford), and the London Midland Region (LMR) also had a London-Birmingham route. The Western had a main line from London to Plymouth and so did the Southern Region. In the case of Birmingham, the LMR route carried more traffic and was to be electrified, while the Western route would carry only a skeleton service of Paddington-Birmingham semi-fast trains. The Western's Snow Hill station in Birmingham would close and all WR services operating beyond Birmingham to Wales, Chester and Birkenhead would cease.

The Western came off rather better where its routes duplicated those of the Southern, although cynics would suggest that it was because SR lines in the West of England were transferred to WR control. Both routes traversed some lengthy tracts of country with little traffic potential. On the SR route, workings were confined to Waterloo-Exeter using WR locomotives. Beyond Exeter the Southern route to Plymouth was severed beyond Meldon. Passenger services operated as a branch line to Okehampton into the 1970s before that service was also withdrawn. The Western retained both its routes to Exeter, the original 'Great Way Round' via Bristol and the shorter route via the Berks & Hants line, though on both routes, numerous wayside stations were closed.

Such changes to the main lines bred further changes of their own. By eliminating wayside stations and reducing the status of lines such as Oxford-Worcester the number of trains was also reduced. By recasting the service it then became possible to reduce lightly-trafficked sections to single track. Turning back the clock to reduce four-track main line to two brought the benefits of simplified track layouts and less complex or expensive installations for the change to multiple-aspect colour light signalling (MAS).

Thus, it is possible to see that the Beeching Report is inexorably intertwined with the whole business of modernisation. Closures and rationalisation would have made radical enough changes to the appearance of the Western Region. But, closures and rationalisation did not come on their own. They facilitated less expensive MAS installations which in turn brought the elimination of manual signalboxes. Without manual boxes, gated level crossings were impractical, and so

Above:
Patney & Chirton, situated on the erstwhile Berks & Hants Extension Railway formed the junction where the original line to Devizes diverged from the Great Western's cut-off route via Castle Cary. The station was thus a GWR original built in the standard style typical of the final years of the 19th century. The Devizes line platform was on the right and beyond the station were extensive sidings. The station is here seen in September 1966, five months after closure of local stations on this section of the main line. Several stations here survived remarkably intact for a considerable time after closure. *Andrew Muckley*

Below:
Another April 1966 closure on the West of England main line was Wootton Rivers halt, seen here on the same date looking towards Newbury. I walked here from Savernake Low Level one hot summer afternoon in 1964 and caught the Westbury-Newbury local service as the first stage of the journey home. Note the new parapet walls on the overline bridge and the concrete posts for the Tilley oil lamps. The halt shelters are an original GWR type but the platform has been rebuilt in either late GWR or early BR period using a standard sectional concrete construction. *Andrew Muckley*

DAUNTSEY

4903
PASSENGERS MUST NOT CROSS THE LINE EXCEPT BY MEANS OF THE FOOTBRIDGE

Top left:
Dauntsey, on the original GWR line to Bristol, between Swindon and Chippenham, was once junction for the branch to Malmesbury. This view towards Swindon in August 1964 shows the original Brunellian stone building and matching waiting shelter. By chance I was on the very last train to call here early one Saturday evening in January 1965. Dauntsey closed together with Christian Malford halt and Wootton Bassett, enabling Beeching's reorganisation of Swindon-Bristol local services to proceed. *Andrew Muckley*

Centre left:
During 1987, in the final months before major resignalling at Newton Abbot, the author was privileged to visit both East and West boxes, as well as the box at Aller Junction. In this view Newton Abbot West signalbox, a substantial wooden structure, is being passed by 'Hall' No 4903 *Astley Hall* on an up express. *Ian Allan Library*

Bottom left:
Newton Abbot East box, with a 206-lever frame was destined to become the largest WR manual box still in existence during the mid-1980s, eventually being swept away in the spring of 1987.
D. B. Hart

Above:
Another fine-looking manual box which survived until 1987 was the brick-built platform-mounted example in standard GWR style at Totnes. With the introduction of MAS in the Beeching era it was inevitable that the westward march of technology would eventually reach even the farthest outposts of the system. *D. B. Hart*

Right:
On the main lines nearer to London, semaphore signalling was an early victim of modernisation, the spread of MAS reaching westwards as far as Reading by 1965. This handsome array of wooden post GWR examples was at the western end of Maidenhead station, where the branch to High Wycombe curves away to the right. A pannier tank is running backwards on the up relief line and the photographer has his back to the fast lines. Note the new and old signalboxes either side of the locomotive, both being replaced by the Slough panel box. The goods shed in the left background survived the withdrawal of freight services under Beeching, and the site was used as a loading area for new cars. *D. B. Hart*

Above:
A more recent view of the branch starter at Maidenhead, looking from the bay platform towards the goods yard. The lower arm serves the route diverging left into the yard and below it is a mechanical route indicator. *Ian Allan Library*

Below:
In this early 1950s view a 'King' 4-6-0 No 6026 *King John* heads the down 'Cornish Riviera' through Taplow. The signalbox is visible on the left with the relief lines passing behind it. To the right is the goods shed which gained some fame as the early base of the Great Western Society after the goods services had closed with Beeching rationalisation. Initially goods services were concentrated on Slough, but in due course the yard there also closed. *R. F. Dearden*

came the lifting barrier, either closed-circuit television (CCTV)-controlled or as an automatic half-barrier installation.

For me, rural wayside stations always had a special fascination. They were tranquil places where, for minutes on end, nothing happened. There was bird song and perhaps the hum of a tractor in a nearby field. As you wandered the platform or sat on one of those comfortable, solid GWR seats there was always an air of expectation. What would come through next, and from which direction? Then the silence was punctuated for a few seconds as an express rushed by, or a lengthy freight rumbled past with its engine wheezing and groaning, and the brake handle of a loose-coupled wagon clanging up and down. The silence returned. A while later a lone railwayman would emerge from his office. The local arrived. There was a banging of doors, some chatter, a whistle, a departure.

Surely one of the most unproductive proposals in the Beeching Report was the desire to 'damp down seasonal traffic peaks' so that it would no longer be necessary to hold spare corridor coaching stock to cover them. In a nutshell this meant summer holiday passenger traffic, plus the extra traffic which was inevitable at Easter time and Bank Holiday weekends. Its seems scarcely credible that any industry should actually propose to discourage customers in this way, even if it did have to carry some under-utilised assets in order to cover these peaks.

As far as the Western Region was concerned, some 'damping down' of demand was bound to occur when lines to holiday resorts were closed. The Beeching Report envisaged Newquay as the only rail-served resort on the Atlantic coast between Weston-super-Mare and Penzance. In Cornwall the St Ives, Fowey and Looe branches were slated for closure (Helston had already closed). With fewer destinations, demand was bound to fall. In practice, some but not all of the West Country branches did close, but post-Beeching, the ever-increasing use of the family car for holidays, and improvements in road access with motorways and by-passes to most of the major traffic black-spots did far more to damp down the demand for rail travel.

The WR had inherited the Great Western's 'Cornish Riviera', the number one English holiday area, promoted and developed over half a century when rail had been the primary means of access. Such was the demand for travel during the holiday season, that passengers on most trains to and from the West of England were required to hold seat regulation tickets. These were a form of advanced booking

which enabled the Great Western, and subsequently the Western Region to assess in advance the level of demand. Extra trains were then laid on as required. A typical holiday train would comprise up to 14 coaches, while principal expresses such as the 'Cornish Riviera' would often require relief services running a few minutes behind them.

From 1951 the WR had done much to continue the promotion of West Country holidays. The 'Cornish Riviera' train was provided with a headboard to make it readily identifiable and a number of other new named trains were introduced. These included the 'Royal Duchy' serving a large number of stations in Cornwall on its through run from Paddington to Penzance, the 'Cornishman' from Wolverhampton to Penzance the 'Devonian' (Bradford-Paignton), the 'Merchant Venturer' (Paddington-Weston-super-Mare), and the 'Mayflower' (Paddington-Plymouth). Holiday destinations in West Wales were catered for by the 'Pembroke Coast Express' (Paddington-Pembroke Dock) and the 'Red Dragon' (Paddington-Carmarthen).

Much of the purpose behind these trains, and their un-named but no less important supporting services, was to promote holiday travel and develop the demand for rail travel. Only a dozen years after their introduction, the Beeching Report was proclaiming that BR did not want such demand because it was uneconomical to keep the coaching stock to cater for the demand. By then, the demand was already declining, but in the late 1960s there came the product of the Beeching Report — a deliberate failure to provide the capacity to cope with whatever demand there was. Thus, traffic was driven away by the simple expedient of not providing adequate services or sufficient seating capacity. After all, a passenger only has to stand in the corridor on one long journey because all the seats are taken, and the next time he decides to use his car.

This aspect of Beeching-style thinking has unfortunately become endemic. With the introduction of fixed formation IC125 units in 1976 the Western Region was left without the flexibility to alter train formations to cope with increased demand. Diagramming and availability are such that reliefs or extras are exceptional while the operation of locomotive-hauled extras means finding coaching stock which is spare and a path in the timetable which does not interfere with the much faster IC125 schedules. In fact, the successful damping down of peak demand, acceptance of large numbers of standing passengers, and the introduction of TOPS diagramming of coaches, has enabled the coaching fleet to be pruned to the extent that hardly any vehicles stand spare for any length of time anywhere on the system.

Perhaps the proposal was right in principle, but as with so many aspects of the Beeching Report, it was the *degree* to which it was applied which causes so much controversy among railway sympathisers.

Above:
Taplow West box was remarkable indeed, being set in a commanding position on top of the footbridge and straddling the island platform. The view along the main lines, particularly the long stretch west of the station must have been quite breathtaking. This box succumbed rather earlier than the East box, as the Western Region rationalised its signalling. *D. B. Hart*

Below:
A well-known location, even from photographs taken in the early years of the present century, was the tall box set between the main and relief lines at Twyford on the edge of Sonning Cutting. *D. B. Hart*

CHAPTER

3 Branch Lines and Byways

The Western Region had more than its fair share of branch lines. Victorian railway fever had affected most rural communities and most towns and villages had at one time or another been the subject of a proposal to build a railway. A substantial number of these schemes came to fruition, but many never envisaged anything grander than a short connecting route to a nearby main line, while others, launched in grand style, petered out without reaching their objective. These sometimes resulted in unfinished through routes or branch lines which, instead of ending at a logical traffic source, ended at the point where the money and the enthusiasm ran out.

They were either rescued from oblivion by the GWR, or, even if financially successful, were eventually swallowed up by their larger neighbour. The Light Railway Act of 1896 aimed at making it easier for rural communities to build railways. Expensive earthworks and other heavy costs could be avoided at the price of accepting a 25mph speed limit. This was scarcely a restriction, since many rural lines offered little prospect of significantly faster running. In the Great Western's area, one or two light railways sprung up. Some, such as the Western, Clevedon & Portishead Railway

Below:
In the twilight years of the branch line, weekend railtours and last workings became a way of life for the dedicated enthusiast. The last steam-worked excursion over the Ashburton branch in 1962 was a brake van tour headed by '45XX' 2-6-2T No 4567 here seen at the terminus. *Ian Allan Library*

Right:
Typical of day-to-day workings over the branch lines of South Devon, a '45XX' 2-6-2T heads a couple of downgraded main line coaches on a Teign Valley line working near Trusham. *Ian Allan Library*

Bottom right:
A panoramic view of the spacious branch terminus at Kingsbridge, with the station building on the left and the locomotive facilities centre, behind the platform. Another of the ubiquitous small Prairies has charge of the branch train formed of a single Hawksworth non-corridor coach and a 'Fruit D' van. *Ian Allan Library*

KINGSBRIDGE

Above:
Beeching poised his 'axe' over the Liskeard-Looe branch, but special circumstances caused this line and the St Ives branch to be reprieved. The 5.50pm Liskeard-Looe is seen entering St Keyne halt on 17 September 1957 behind '4575' class 2-6-2T No 5557. The service was dieselised in the early 1960s with single-unit railcars and nearly 30 years on the same vehicles are in use, more modern stock being unsuitable for the line's sharp curves and steep gradients.
Ian Allan Library

(WC&PR) and the Shropshire & Montgomeryshire were part of the Col Stephens empire of ramshackle light railways. The WC&PR did come into GWR control in its twilight years, while there were others such as the Wrington Vale Light Railway which lasted right into the BR era, albeit as a freight-only line.

The branch line lifestyle was an enduring one. It was essentially a product of the staff and the people who used the line. The idiosyncratic portrayal of staff and passengers in the film 'The Titfield Thunderbolt' was by no means far from the truth during the first 15 years of the BR era. Many is the tale I have heard of trains returning to pick up a late passenger, a driver stopping to pick up a pheasant he had just run down, trains stopping for crews to pick mushrooms — not to mention a signalman who used his level crossing gates for firewood after he had failed to open them for an approaching train!

Dieselisation made little difference, either. I recall two incidents involving the Tetbury branch, a line which I loved and shall refer to more than once in the coming pages. I knew it only in the days of the little AC Cars railbuses, and

Above right:
Beeching closures ended the usefulness of a number of through routes. Closure of the High Wycombe-Bourne End section of the erstwhile Wycombe Railway did not occur until 1970 and was a belated effect of Beeching-style policy. It turned a through route into a truncated branch and ended the prospect of such workings as this Leeds-Windsor & Eton Central excursion headed throughout by 'Jubilee' 4-6-0 No 45562 *Alberta*, seen passing Bourne End on 2 May 1963. Beeching rationalisation reduced the WR station in Windsor to the point where it could not accommodate anything other than the DMU which shuttles back and forth to Slough. *G. T. Robinson*

Right:
By no means all the 1960s closures affected rural areas; there were urban branch lines condemned by Beeching, too. A Pressed Steel single-unit railcar from West Drayton-Staines West dashes past Colnbrook Estate halt in November 1964. The halt had been built only three years earlier in anticipation of a healthy traffic to and from local factories, but in fact only morning and evening rush-hour services usually needed to stop there. *Andrew Muckley*

45562

COLNBROOK ESTATE
HALT
2A82

fascinating vehicles they were, too. The timetable appeared to show a lengthy layover at Tetbury after each trip down the branch. On our first visit we reached Tetbury around lunch time and were rather alarmed when we saw the railbus departing towards Kemble only moments after we had alighted from it. A walk down the road in the sunshine revealed the railbus parked at Troublehouse halt, a wooden platform in the middle of fields. There was no sign of the crew, but the only nearby habitation was the Troublehouse Inn, so we drew our own conclusions!

On another occasion we were concerned about making our connection on to the main line train at Kemble. As so often happened at that time, the timetable showed either very tight connections, or worse still, potential connections which seemed deliberately arranged to 'miss' by a couple of minutes. I enquired of the driver whether or not we were likely to make our connection. His reply was that the trains were not meant to connect, but they always did. In fact, the railbus was late leaving Tetbury but still made the connection with time to spare!

I guess we all had our favourite branch lines, and Tetbury was certainly high on my list. Another was my local branch, from Staines West to West Drayton. Though deep in the WR's London District and operated by stock and crews from Southall, this remained a rural branch to the end. True, there was industry and private sidings at the West Drayton end, and around the intermediate station at Colnbrook. South of Colnbrook, however, the line traversed Staines Moor, a wild and remote area of Green Belt, now scheduled as an area of special scientific interest.

It was an oasis of rural scenery whose peace was broken only by the aircraft noise from Heathrow — which was not as constant in those days. Bisected by the rivers Colne and Wyrardisbury, it was a place where cattle grazed and children fished — for ugly things we called 'bullheads'. About hourly the railcar would amble down from Colnbrook, climbing up the long embankment to cross over the Southern line and drop down into Staines West station.

Right:
Token exchange at Colnbrook through the cab window of a Pressed Steel railcar. The signalman has just received the wooden train staff for the section from Staines West and the driver's right hand holds the alloy token for the section onwards to West Drayton. Colnbrook signalbox controlled one of the last GWR gated crossings to survive in the London area and was eventually demolished after a train wrecked the gates and modern 'open crossing' arrangements were introduced. *Andrew Muckley*

Bottom right:
The railcar from Staines West has arrived in the bay at West Drayton. The wooden shed was the auxiliary token hut which contained a token machine, enabling the token arrangements to be operated without the need to take the token to the signalbox each time. Grand nameboards like this one, which gave details of connecting branch services were progressively painted out and then removed as Beeching closures took place. Nowadays such informative nameboards are rare indeed.

Below:
Photographed when new to the Staines branch in 1958, a Gloucester single-unit railcar pauses for custom at Poyle halt which served the isolated community of Stanwell Moor, on the western edge of Heathrow Airport. This was a GWR halt dating from the 1920s. By the time of its closure in 1965, vandals had burnt down the shelter and the halt was just a bare platform. The Gloucester railcars worked in the London area for only two years before arrival of similar Pressed Steel units. *Author's collection*

Here, the goods yard was disused and overgrown, but the remains of the connection into Staines Linoleum works could still be seen. On the other side of the line the conical water tank and a 'Toad' brake van body grounded as a bothy stood sentinel over the pit, all that remained of the engine shed. Nowadays one of the few clues to its existence is a pair of houses called Great Western Cottages, from where it is said one of the locomotive crew used to climb over the fence at the end of his garden to go and light up the locomotive.

The branch closed in 1965 but thanks to the odd railtour it was still possible to travel over it occasionally for another 16 years, for it was one of the few lines which remained open for freight traffic, in this case central heating fuel oil. Although the Southern end of the line did remain largely rural in appearance, encroaching gravel workings on one side and a huge new reservoir on the other substantially altered its character. It finally succumbed in 1981 when it was severed at an oblique angle by the M25 motorway works, although much of the trackbed is now a cycle and footpath.

A trip up the branch made the perfect prelude to a day out, whether it be train-watching at Iver, or perhaps going further afield. The station was 15min walk from home, and the Pressed Steel single-unit diesel railcar always seemed comfortable and warm and was waiting there when you arrived. An early morning departure might mean catching the three-car unit which formed the one through working (weekdays only) to Paddington. Even in those days, when they were quite new, the DMUs vibrated and the droplight windows rattled in time to the gentle rolling motion as the train headed up the bank out of the station. The thick growth of willows along the line near the Wyrardisbury river would be white with frost, as would the wooden sleepers.

3A26

2A82
WEST DRAYTON
& YIEWSLEY

Left:
As Beeching sought to close the Staines branch to passengers, so freight traffic in the form of oil tank trains was reintroduced, the last freight trains having reached Staines West in 1955. Here, a 'Hymek' Type 3 runs round its train of Shell/BP four-wheelers containing central heating oil, before propelling them into the depot siding. From 1965 until the M25 motorway was built in 1981, the branch from West Drayton remained *in situ* for this infrequent service which ran only twice a week in winter and barely once a month in summer. *T. Wright*

Below left:
Though regular branch services had warranted no more than a single railcar, the final day's services on the Staines branch produced a three-car unit. The very last train, just after dark on an early spring evening in 1965 drew large crowds and the platform at Staines West was packed as a local newspaper photographer caught this view, with detonators exploding under the wheels. *Middlesex Chronicle*

On a wet day overhanging willows would slap the front windows of the train, leaving wet leaves tangled in the wipers. There would be a brief call at Poyle halt, but Poyle Estate was usually passed without stopping. Immediately before Colnbrook level crossing a home signal was passed and the train slowed to a crawl so that the single line tokens could be exchanged through the signalbox window. Colnbrook was the only passing place on the branch, the section from Staines being controlled by a wooden train staff, while that to West Drayton had an alloy electric token. After the token exchange the train would come to a halt in the up platform. The long passing loop was a legacy of wartime use. Once beyond the loop the train rattled under the Bath Road and through Colnbrook Estate halt, usually without stopping.

A tarmacadam coating plant and scrapyard on the right of the line heralded Thorney Mill yard, part of West Drayton's goods yard complex, and then the branch passed under the main line to swing round sharply and steeply to the right into West Drayton station. Shortly before entering the platform, the remains of the branch to Uxbridge trailed in from the left. This line had closed a couple of years before Beeching, Uxbridge having other and better rail connections by way of London Transport.

Pre-Beeching closures had left other freight-only lines, even in the early 1960s. One which has already been mentioned was the Craven Arms-Much Wenlock line. Although closed to passengers, the section from Much Wenlock to Longville remained in use by occasional freight trains until Much Wenlock station itself was closed and the whole line taken up. The traffic comprised mainly agricultural feeds and fertilisers to Longville and occasionally produce in the opposite direction.

Another branch which retained a freight service over part of its length was from Newbury-Lambourn. Though most well known as the Lambourn Valley Tramway, which was

Below:
One of the most curious closures which followed on from Beeching was the decision to close the Bourne End-High Wycombe section, leaving Maidenhead-Bourne End as a truncated branch, with the branch to Marlow trailing in to Bourne End in the wrong direction! This view from the footbridge at Bourne End has changed out of all recognition, for there are now stop blocks and a platform in front of the road, the crossing has gone and everything beyond it has disappeared under redevelopment. This is an early 1960s view – note that the crossing beyond the loop points has already been converted to lifting barriers. *Ian Allan Library*

Bourne End station, looking towards Maidenhead, with a North British Type 2 waiting in the siding with a northbound freight. The branch line to Marlow takes the curve round to the right, and was once operated as a separate branch with autotrains working from the bay platform on the right. With the introduction of DMUs, certain Maidenhead-High Wycombe workings would include a 'side-trip' to Marlow and back before continuing northwards. *Ian Allan Library*

how it started life, the branch became quite heavily used in GWR days for race horse traffic from stables on the Lambourn Downs. After passenger services were withdrawn, the Newbury-Welford Park section was retained for occasional traffic to RAF Welford, in practice a US Air Force munitions store. Type 2 diesels worked the trains of explosives, every pair of loaded wagons separated by several 'barrier' vehicles.

My brother and I walked the trackbed from Lambourn to Welford one wet summer day, and there was a final passenger train when the section to Welford Park finally succumbed in the 1970s. Didcot Railway Centre has the shelter from Stockcross & Bagnor halt and the ticket office from Welford Park, and has produced a good likeness of a typical Lambourn Valley branch halt.

Further west, there remained a short section of the old Midland & South Western Junction line, from Savernake through to Marlborough. It was used for a short while by school specials to and from Marlborough College. The branch to Wallingford was retained for traffic to a malting factory. Around Kington and Presteigne a little network of branches survived for a while, used occasionally for agricultural traffic. Some lines in the Forest of Dean, and the ex-Midland branch to Nailsworth, were retained for coal traffic. In the majority of cases, though, where the track was left *in situ* it was for the purpose of serving one particular customer. In very few instances did a general goods service outlast the passenger services.

All too often the announcement of withdrawal of passenger services meant total closure. There would be a final fling on the last day of service, with dignitaries in attendance and often more passengers than the line's routine passenger train could cope with. After the crowds had gone and the buildings had been locked for the last time there followed only dereliction and decay.

Above:
Another truncated through route was Worcester-Bromyard-Leominster, closed beyond Bromyard as early as 1952, and completely by Beeching in September 1964. Just a month before closure the Bromyard terminus had been reduced to just half a platform in use, and was served by a Gloucester single-unit railcar. *Andrew Muckley*

There would then be a visit by a lorry, or perhaps a train, listed in the working notices as 'recovering redundancies'. This usually meant removing seats, platform trollies, nameboards, furniture and the like, and would happen quickly after the closure. Most stations then suffered a period of dereliction and sometimes it might be a year or more before the track was lifted. When the track went, wooden buildings were usually demolished and burnt, while more substantial structures were left to the attentions of vandals before the site finally changed hands for either demolition or conversion.

Just as one or two main line wayside stations escaped closure when Ministerial consent was not given, so one or two branches also lived to fight another day. One could have hoped for more. The main successes were in Cornwall and showed the level of exceptional circumstances which were required if the Minister were not to give his consent to closure. Of the three Cornish branches proposed to lose their

503
B

Above left:
Next station from Bromyard in the Worcester direction was Suckley, a rarely photographed and rather unusual building of board-and-batten construction. The standard GWR shelter on the opposite platform is of more recent origin. *Andrew Muckley*

Left
A Gloucester three-car cross-country unit pauses for custom, including a bicycle, at the remote Knightwick station on the Worcester-Bromyard branch in 1962. *Andrew Muckley*

Above:
One of the shortest and busiest WR branches, even to this day is that from Slough to Windsor & Eton Central. However, since the 1960s it has been progressively rationalised, reduced to single track, cut from five platforms to one at the terminus and with the west curve at Slough removed, preventing direct access to the branch from the Reading direction. In happier days, 'County' 4-6-0 No 1014 *County of Glamorgan* heads an excursion out of Windsor over the River Thames bridge, the earliest surviving example of a Brunel wrought iron bridge. It is running on the up line which was removed in the early 1960s.
P. Howland

Centre right:
The Shrewsbury-Hartlebury Severn Valley line ran through the Severn Gorge and at Ironbridge & Broseley the station was close by Abraham Darby's iron bridge. A southbound Severn Valley train pauses at Ironbridge with ex-LMS 2-6-2T No 41209 in charge. This section closed in September 1963 and by the time of this photograph one platform was already out of use. *Andrew Muckley*

Right.
At Buildwas Junction the Severn Valley line was crossed by the branch from Wellington to Much Wenlock. The branch climbed at 1 in 40 and was already higher than the other line when it reached Buildwas station, resulting in this curious arrangement. This is the view towards Shrewsbury, with the branch platform behind the fence on the left, on 24 August 1963, shortly before closure. Once closed, this station disappeared very rapidly under the site works for the new Ironbridge 'B' power station. *Andrew Muckley*

BUILDWAS JUNCTION
WELLINGTON LINES

Left:
In its final months the Shrewsbury-Hartlebury service was worked by DMUs and one of the ubiquitous Gloucester single-units usually sufficed. Bound for Shrewsbury, the railcar calls at Buildwas on 24 August 1963. Ironbridge 'A' power station is visible on the left and was the main reason why this isolated rural junction station had a vast and complex track layout controlled by a signalbox with over 100 levers. *Andrew Muckley*

Below left:
Inside Berrington signalbox on the northern section of the Severn Valley line. This was a remote rural station and crossing place on the single track route. Note the token instrument and pick-up hoops behind the signalman. *Andrew Muckley*

Above right:
South of Buildwas the line also became much more rural as shown by the next few photographs. The railcar lurches and sways into Linley station, typical of the standardised architecture on the line.
Andrew Muckley

Right:
The Highley signalman has walked over from his box and waits with the single line token for the next section as the Shrewsbury-bound railcar arrives. The footbridge provided a slightly shorter route towards the village but was removed when this section of the route was sold to the Severn Valley Railway for preservation. *Andrew Muckley*

Above left:
A Derby three-car unit bound for Bewdley pauses at Arley, running wrong line through the station on 25 August 1963. A notice on the platform indicates the prize-winning length of track in a time-honoured competition which helped to give permanent way staff (or gangers as they were in the good old days) some job satisfaction. *Andrew Muckley*

Left:
Post-Beeching, the Bewdley-Highley section remained in place for occasional traffic from Alveley colliery, and so the story goes, it was occasionally used by a train carrying nothing more than the wage packets for the staff! The scene at Arley in May 1972 shows the track slewed away from derelict platforms. The Severn Valley Railway has since restored this station to pristine condition. *Andrew Muckley*

Above:
Kidderminster's lovely half-timbered station was said to have been a kit of prefabricated parts intended for Stratford-upon-Avon but diverted to Kidderminster when a fire destroyed the station there. Dry rot was discovered in the building in the 1960s and it was demolished with indecent haste. *Andrew Muckley*

passenger services Lostwithiel-Fowey was closed. Here it had been suggested that the passenger service was not viable, and more importantly, the operation of passenger trains interfered with the running of the china clay traffic for which Fowey was, and still is, the principal port.

In the autumn of 1966 the Labour Government's Transport Minister, Mrs Barbara Castle, refused consent to the closure of the St Erth-St Ives and Liskeard-Looe branches on the grounds that the local roads were inadequate for the additional traffic. It was welcome news but both branches were rationalised to the point where staff and even station buildings were removed. In due course, both branches were cut back at the termini to provide extra car parking space. Latterly an enterprising 'park and ride' scheme was introduced at St Ives to reduce traffic congestion in the town. Visitors can now park at a new halt at Lelant Saltings, some three miles outside St Ives and use the rail service into the holiday resort. Such has been the success of the scheme that summer trains on the branch are now formed of four cars (six on occasions), while a single diesel railcar sufficed in 1966. Efforts to persuade visitors to Looe to park at Liskeard and use the train have not been marketed so vigorously and have not met with quite the same degree of success.

The St Ives branch has certainly shown that in some cases there were better cures than the medicine proposed by Beeching. Fowey faces similar traffic problems to the other two Cornish resorts and could have benefited from a similar scheme. However, the station site at Fowey has been flattened to serve as a car park and there is still the problem of the heavy china clay traffic on the branch.

The former SR line from Barnstaple to Ilfracombe closed in 1970, a comparatively late effect of Beeching. Thus only two lines, both reduced to very basic branch status, Exeter-Barnstaple and Par-Newquay of considerable length cross the grain of the West Country in a general southeast-northwest direction. They each muster only some half dozen trains per day and would doubtless be on the 'hit list' of any modern-day Beeching. Yet, 25 years after

One of the most popular and picturesque of WR cross-country lines was the Gloucester-Hereford route through Ross-on-Wye. 'Manor' 4-6-0 No 7815 *Fritwell Manor* rolls a Gloucester-bound train through Weston-under-Penyard halt in August 1964. The platform surface would not have passed modern safety inspections. *Andrew Muckley*

Above:
Ross-on-Wye station, built in 1892 to a design by J. E. Danks, marked a turning point in Great Western architecture with new standard styling below the eaves, and traditional roof turrets with spiky ironwork. The result was very fine and it is no surprise that the new Severn Valley Railway station at Kidderminster is modelled on this design.
Andrew Muckley

Left:
A West Country branch which escaped closure and which has since gone from strength to strength is the 4½ miles from St Erth to St Ives. By the summer of 1966 Beeching's rationalisation had already had its effect, with the track in the platform cut back and all other lines removed. In due course the buildings were demolished and the platform moved in order to make the maximum possible area available for lucrative car parking. *Author*

Above:
A West Country branch whose passenger services did not survive served the picturesque port of Fowey. Here it was said that the lightly-used passenger service caused problems for the extensive china clay traffic. The branch was truncated at the china clay berths and the station area is now a car park. In August 1964 the branch railcar had been shoved in a siding (right) to clear the platform road for a freight working. China clay hoods on the left await movement to the docks for unloading. Old coaches were often provided as offices and stores at busy locations. The Fowey branch passenger service ceased in January 1965, one more victim of Beeching. *Andrew Muckley*

Beeching, the lessons have not been learned. Instead of positive promotion and development to attract travellers, there is only a train service pared down to a bare minimum. The Barnstaple line offers bad or non-existent connections with main line services at Exeter, while the Newquay line, blessed with through IC125 units to Paddington on Saturdays has them running almost non-stop between Par and Newquay, thereby depriving the line of its local service on the very day when local people might be persuaded to use it.

Below:
The increasing pace of branch line closures and demolitions encouraged numerous railtours over lines which had long since lost their passenger services and were freight-only. One such was the Moreton-in-Marsh to Shipston-on-Stour branch, closed to passengers in July 1929. Only the lightest locomotives were permitted over this little-used line and both a 'Dean Goods' 0-6-0 and one of the ex-Midland & South Western Junction 2-4-0s worked specials to Shipston. One of these is seen at the terminus in August 1952.
Real Photographs (K1634)

CHAPTER 4

Freight — Main Line and Branch

The observer watching trains on today's Western Region would be hard put to imagine the scale and diversity of freight operations which existed on the railway before Beeching. Today's freight trains are largely a product of the Beeching era, for the main aim of the Report as far as freight was concerned, was to have the railways carry only those commodities which could be carried profitably. Thus the requirement was for block trains, complete trains carrying a single commodity from one place to another with no pick-ups or drop-offs en route.

So, today's freights carry coal, oil, iron ore, steel products and the like. It is doubtful whether Beeching foresaw the huge growth in aggregates traffic which now constitutes a large part of Western Region freight traffic. The Beeching proposal to experiment with 'liner trains' of course resulted in Freightliner and the development of the very successful rail-borne container operations. However, the role of Speedlink operations in enabling siding-to-siding traffic with individual wagons has probably developed beyond anything envisaged by Beeching.

There is no doubt that large amounts of the freight traffic carried by the Western Region in the first two decades of BR was being carried unprofitably. In part this was due to the railway system being incorporated as a 'common carrier'. This meant it was obliged to convey any traffic offering, however impractical it might be. Thus, all sorts of special wagons were required for such things as cattle, horses, fruit, fish, and machinery. One of the last of these special wagon types to be phased out was the end-tipping, wooden-bodied china clay wagon peculiar to the Western Region. These lasted in traffic until 1988.

So, Beeching sought to rid the railway of its common carrier status to enable it to pick and choose the traffics which it was best suited to handle — large bulk loads which would constitute complete trains. This would enable thousands of expensive and often elderly wagons to be scrapped. By getting improvements in the coal mine-to-power station

Left:
'57XX' 0-6-0PT No 3796 approaches Newton Abbot on 28 July 1962 with freight from Bovey Tracey and Heathfield on the former Moretonhampstead branch. *Ian Allan Library*

Below:
The same locomotive comes to a stand outside Newton Abbot East signalbox as the fireman passes down the single line token for the branch to the signalman. The first two vehicles are banana vans, representing one of many types of traffic lost to the railway in the wake of Beeching. *Ian Allan Library*

'2884' class 2-8-0 No 3812 heads a typical mixed freight including vans, open wagons and containers, on the South Devon main line.
Ian Allan Library

Above:
The Teign Valley line from Exeter-Heathfield closed to passengers in June 1958. Newton Abbot-based '41XX' 2-6-2T No 4148 trundles the branch freight through the remains of Trusham station on 17 March 1962. *Ian Allan Library*

Left:
At Teignbridge Crossing near Newton Abbot, an 0-6-0PT of the '57XX' class has charge of a lengthy freight, emphasising the usefulness of such branch lines in keeping numerous lorry movements off congested roads. The crossing keeper's cottage and ground frame hut were to become anachronisms in the Beeching era. *Ian Allan Library*

Left:
Dulverton station on 14 September 1953. The Exe Valley line autotrain from Exeter formed by ex-railmotor coach No W9 is arriving behind '14XX' 0-4-2T No 1429, still in Great Western livery. At the main platform a Taunton-Barnstaple train is also arriving, with a '43XX' 2-6-0 in charge. In the foreground three cattle wagons await their 'passengers', another source of freight revenue viewed as uneconomic by Beeching. *Ian Allan Library*

4174

Left:
'41XX' 2-6-2T No 4174 has arrived at Torquay gasworks with a train of coal from Kingswear on a sunny day in 1962. The Great Western usually tried to avoid complex trackwork such as the double slip visible ahead of the locomotive. *Ian Allan Library*

Below left:
A '45XX' 2-6-2T shunts an ex-GWR 'Toad' brake van in the massive South Devon railway timber goods shed at Totnes. At the other end a North British Type 2 diesel-hydraulic has arrived with a mixed freight. This sort of pick-up freight working to wayside stations came to an end with the Beeching Report, and what appears to be a service withdrawal poster is visible through the right-hand shed door. *Ian Allan Library*

operation, and by concentrating domestic coal operations on a few specially equipped 'concentration yards', similarly large numbers of old coal wagons, the vast majority without continuous brakes, could also be disposed of.

In fact, where coal was concerned, other developments helped to speed the project. The simultaneous spread of smokeless zones under the Clean Air Act and the increasing adoption of central heating, fuelled by gas, electricity or oil, reduced the demand for domestic coal. Local coal merchants were in decline, either being absorbed by larger companies or transferring to other aspects of the domestic heating industry. So, just as passenger stations had seen an accelerating pace of closures through the 1950s and early 1960s, so too did wayside coal yards. Often, as coal constituted the lion's share of the local freight traffic, closure of a coal yard could mean that a complete goods yard might close or a complete local freight service be withdrawn. Yet, particularly at more rural locations where the move away from coal was slower and there was little need for smokeless zones, coal traffic was often the last freight traffic to survive.

Coal's major customer was the electricity generating industry, and here the railway required, and got, the co-operation of its two sister nationalised industries, the National Coal Board and the Central Electricity Generating Board, to develop a new concept in conveying coal from mine to power station. On the Western Region, coal from South Wales was traditionally moved in block trains of unbraked mineral wagons. The only major development in half a century or more had been the introduction in the 1920s of a steel-bodied, long-wheelbase four-wheeled wagon of greater carrying capacity. These were the so-called 'Felix Pole' wagons, after the GWR's General Manager of the time. By the 1950s many were life-expired and were replaced by the BR standard 16-ton mineral wagon with no major technical advance, and less capacity.

With Beeching came the concept of a railway 'conveyor belt' continuously moving block trains of coal from mine to power station and returning empties. New loading facilities were required at mines, but pit closures and concentration of resources on a smaller number of high-productivity mines made this practicable. Similarly, construction or reconstruction of power stations meant that these could be equipped to receive the new trains. From the railway viewpoint, key to the whole operation was complete re-equipment of the coal wagon fleet with new high-capacity wagons equipped for hopper-unloading through bottom doors. This was to become known as the merry-go-round (MGR) system.

Right through to the end of Western Region steam in the mid-1960s, the changes in freight operation were almost imperceptible. To the casual observer it was only the number of freight trains which gradually diminished. Perhaps, because of the loss of one particular traffic or another, the surviving freights became slightly shorter. There were clues to declining freight operations, though. Taking, for instance, my home locality; Staines West goods yard had closed in 1955 with alternative facilities available at Staines Central (SR) or Colnbrook. When Colnbrook yard closed, alternative facilities were offered at Slough. West Drayton yard had by then become one of the first coal concentration depots, and remains so to this day. Such was the progress of declining freight traffic that, long after Beeching, even Slough goods yard closed. A large town, with substantial industry, was unable to support even a Speedlink siding.

There can be little doubt that much of the loss of freight traffic on the railways can be laid at the door of the negative outlook reflected in the Beeching Report. It saw no future for rail freight other than in bulk commodities, and failed to

Above right:
Intermediate block signalboxes were provided at Shottesbrooke (seen here) and Ruscombe in order to break up the long section between Maidenhead and Twyford and so increase the line capacity. This was particularly important for wartime traffic and also for peak summer Saturdays, when despite these provisions the line would be used to full capacity. Inevitably the need for them declined as traffic fell off and with the spread of MAS they were early casualties. A '28XX' 2-8-0 No 2835 tramps down the relief line at Shottesbrooke with empty ballast hoppers for Tintern quarry. *Ian Allan Library*

Right:
The big '47XX' 2-8-0s were built for fast freight traffic and No 4701 is here seen wheeling a typical milk train along the main line west of Maidenhead. Even the Beeching Report did not envisage the end of milk traffic by rail, but this too has happened. *Ian Allan Library*

offer effective, aggressive marketing approaches at the very time when the road transport industry was getting ever stronger. Two glaring illustrations are regularly seen in southern England. Early morning on the M3 motorway brings a steady stream of banana lorries heading from Southampton towards London. In the early 1960s the banana trains were a regular part of the rail traffic past my home. Equally, the motorist using the A30 or A303 roads is sure to find himself caught up behind a milk tank lorry at some stage. It is doubtful if even Beeching envisaged that this traffic would also be lost to rail.

Rail-borne milk traffic was part of the Western scene in the 1960s, there being daily milk trains converging on London by way of all the main line routes. Some, though not all, of this milk traffic began its journey on rural branch lines. In GWR days most country stations had offered some milk traffic in churns. The change to tank wagons in the 1920s and 1930s led the traffic to be concentrated at certain creameries provided with tank-filling equipment. An example where milk traffic outlasted the passenger service was the Tiverton-Hemyock branch. Milk loaded at Hemyock was tripped to Exeter where tanks from Torrington would be added before attachment to the main train for Kensington which had originated at St Erth. In steam days a 'Castle' 4-6-0 was the usual motive power, but by the 1960s the job had passed to 'Warship', and later 'Western' class diesels. Cessation of the Hemyock and Torrington services coincided with the rundown of milk movement by rail in the late 1970s. Eventually the traffic was transferred to road, with only a handful of newly converted milk tank wagons kept in reserve which were finally put up for disposal in 1989. Withdrawal meant the end of the Hemyock branch, while the line to Torrington lasted only a little longer.

The line from Barnstaple through Torrington to Halwill Junction had become part of Western Region along with the other SR lines in the West of England, concurrently with the implementation of the Beeching Plan. Passenger services beyond Barnstaple (except to Ilfracombe) ceased in 1966, but milk from and fertiliser traffic to Torrington remained, together with china clay emanating from Meeth, a few miles beyond Torrington. This section of the line had actually been opened as late as 1925, engineered by Col Stephens under Light Railway regulations. It had never offered much passenger traffic, most services being one coach mixed trains with goods wagons attached.

Although Meeth continued to produce china clay and ship it by rail into the early 1980s, it was the line's light railway origin which was its undoing. With the pressing need to replace the life-expired wooden-bodied china clay wagons,

Below:
Milk trains originated at the extremities of the system, including Whitland and St Erth. In this view 'Grange' 4-6-0 No 6824 *Ashley Grange* has collected tanks from the yard at St Erth and is reversing out on to the up main line. The train will make other stops including a locomotive change and a stop to pick up the Hemyock tanks at Exeter on its way to Kensington. *P. Q. Treloar*

Right:
Running under a classic 'droopy' signal, a '28XX' 2-8-0 heads a heavy coal train down the South Devon main line. Once the staple traffic from South Wales, the railway's coal traffic declined with the mining industry in that area. The third vehicle in the train is evidence that new 16ton mineral wagons were still being built, moving coal in the same way it had been moved for more than 50 years. With Beeching came the revolution in hopper wagons and MGR operations. *Ian Allan Library*

various private owner wagon companies offered bogie covered hopper wagons of high capacity, suitable for movement in Speedlink freight services. However, the track and the light railway infrastructure beyond Torrington proved unsuitable for the new vehicles and so the china clay was moved by road to suitable rail loading points. With its final traffic source gone, the railway beyond Barnstaple closed completely in 1981 and much of the trackbed is now incorporated in the coastal footpath.

One of the more surprising features of the modern railway is the way that some WR main lines have lost all their freight traffic. The Oxford-Worcester line, the 'Old Worse & Worse', now sees no regular freight traffic, and there is little or nothing over the Gloucester-Swindon route. The Berks & Hants line might have gone the same way were it not for the establishment, post-Beeching, of huge rail-served quarries near Frome and Shepton Mallet. A freight depot, built on the site of the Permanent Way Pre-assembly Depot at Theale receives oil, cement and aggregates but otherwise the only freight over the line is occasional domestic coal, and supplies for the MoD establishment at Thatcham.

Yet, the WR once carried so much freight. On a summer Saturday's train spotting at Iver in August 1962 I logged 15-20 freights out of around 80 services noted, and Saturdays were not the best days for seeing freight trains. The numbers of holiday extras which ran on a Saturday precluded some freight movements, so on a weekday one could reckon that 25% of the services seen on the four-track main line at Iver would be freights. Most of them, at this particular location, would be mixed freights, there being comparatively little in the way of block train traffic at this point. Indeed, such was the volume of local freight — trip working — between the yards at Slough and West Drayton that a goods-only relief road was provided. This long loop on the north side of the up relief line extended from a point west of Iver, past Iver sidings, to West Drayton yard. It survived the Beeching era of removing redundant track and remains in place today, now linked right through the former Staines branch bay platform at West Drayton, to join the up relief line just east of the station. It still sees a modest amount of usage, though nowadays it is usually to recess lengthy freights to allow passenger services to overtake.

The Beeching Report sought wherever possible to eliminate duplication of effort. In taking over SR lines in Devon and Cornwall the WR had acquired a second quarry from which to extract track ballast, at Meldon near Okehampton. Most of the WR's ballast had come from its own quarry at Tintern, and I well remember watching 'Hall' 4-6-0s and '28XX' 2-8-0s attack Sapperton Bank with their long trains of the surprisingly small 'Dogfish' ballast hopper wagons, banked in the rear by a 2-6-2T from Brimscombe. Subsequently, all WR ballast quarrying was centred on Meldon.

The system of wagon forwarding and the complex movement of an individual wagon to its destination often involved frequent remarshalling en route. This was expensive in time, manpower and facilities. The sheer number of wagons and the multitude of different destinations would have precluded anything but the most laborious of manual systems. It is hard to imagine a TOPS-style computer system being able to cope with that scale of operation.

KINGS SUTTON
3820

Left:
At the other end of the coal delivery scale was the local branch freight still delivering to coal merchants via a branch line which had closed to all other traffic. Swindon-built Type 1 0-6-0 No D9527 shunts the Nailsworth branch freight at Dudbridge on 3 January 1966. This former Midland Railway branch was deep in the heart of GWR territory and so became part of the Western Region. *Andrew Muckley*

Below left:
Enjoyable Saturday afternoons could be spent at the lovely old country station at Kings Sutton, near Banbury. It was notable for its wildlife, and common lizards (quite rare in England despite their name) were to be found in the rockery beneath the nameboard. In gleaming ex-works condition '2884' class 2-8-0 No 3820 heads an up mixed freight bound for Oxford in August 1962. *Andrew Muckley*

So, instead of taking the wagon right to the customer, or to his nearest goods yard, the customer was expected to take his goods to the wagon at the appropriate freight railhead. Once on the lorry, of course, it made sense for the customer to take his goods right through to their destination by road and thereby save the time and risk of damage during transhipment. By cutting out the short local trips at each end of the wagon's journey and making these movements by road, remarshalling of the wagon was avoided. With hundreds of such remarshallings eliminated, small marshalling yards could be closed. Freight marshalling movements were then concentrated on a few large yards in the major conurbations, a process of rationalisation which had begun even before Beeching.

The frequency of line closures gathered momentum through the 1950s and reached its peak with the passenger service withdrawals made by Beeching. Branch line service withdrawals were sometimes a haphazard affair which cannot always have resulted in the desired economies. The 1950s policy of 'pruning' only the worst offenders tended to have the effect of reducing traffic and increasing costs on adjacent and connecting lines which might already be heavy loss-makers.

In some cases the cessation of passenger services left only an occasional freight working over a particular route. Sometimes these ran only on an 'as required' basis. In these circumstances there could be no economy from removal of track or sale of land, and some staff might even have to be retained.

Although Beeching era withdrawals removed some of these freight-only lines, a whole rash of new ones were created on the WR in the 1960s. For various reasons, few survived the quarter century after Beeching, though some operated for at least a decade.

Having become an ardent WR fan in the 1960s I usually made a point of beginning my train journeys at Staines West with the 15min journey up the branch to West Drayton. When I first got to know Staines West the goods yard was derelict and overgrown. All the buildings and sidings were still there although the signalbox and the connection from the yard to the running line had been removed, freight facilities having been withdrawn in the mid-1950s. Before that the traffic had been mainly coal to local coal merchants, although there was also a connection into the Staines Linoleum Company's works.

Linoleum floor covering had been invented on the Staines site in the 19th century and although nowadays superseded by vinyl, it had enjoyed spectacular popularity for the first half of the 20th century. The process of manufacture used hemp and resins brought in by rail and in the early years of the present century the GWR had actually provided two special wagons for the conveyance of linoleum traffic between Staines and Smithfield Market. By the 1960s the factory had ceased to send goods by rail and its Staines manufacturing operation was in decline.

Between Staines and West Drayton the only crossing place was at Colnbrook, where there was a goods yard served by a daily freight working from West Drayton. North of Colnbrook there were private sidings at Colnbrook Estate halt and Thorney Mill. Traffic was mainly steel for factories and particularly reinforcing rods for use in precast concrete manufacturing. Thorney Mill yard flourished after the West Drayton goods yard was taken over entirely for the first of the Beeching era Coal Concentration Depots built for the distribution (by road) of domestic coal. At Thorney Mill the yard received tar for use in road coatings, and despatched scrap metal from a large scrapyard. At least two BR Standard '4MT' 2-6-4Ts were cut up there, but the scrap merchant dealt mainly with cars and machinery.

An early morning departure from Staines West would usually mean that we passed the branch goods shunting at Thorney Mill. It was usually a '57XX' 0-6-0PT accompanied by the inevitable 'Toad' brake van branded for use on the branch.

At West Drayton, the single branch platform on the north side of the line served both the branches to Staines West and Uxbridge. Here would wait identical railcars, one for each destination, a practice which caused perpetual confusion and not a few wasted journeys down the wrong line. Despite its more densely populated catchment area, the Uxbridge line lost its passenger service in 1962 — before Beeching, and three years before the Staines branch. My spotting notebook for both 10 and 17 August 1962, a month before the Uxbridge line closed, shows '61XX' 2-6-2T No 6135 working a three-vehicle afternoon parcels train over the branch.

Such was the apparent lack of co-ordination that, even as the notices announcing withdrawal of the Staines branch passenger service were being posted, the derelict goods yard at Staines West was being redeveloped as a rail-served depot for cental heating oil. '61XX' 2-6-2T No 6143 worked the first oil train into the depot in July 1964 and the passenger service ceased in March 1965. So, the track south of Colnbrook still had to be retained for a service which ran about once a week in winter and once a month in summer. Apart from the very first train all subsequent workings were diesel-hauled, usually in the early days by a 'Hymek' Type 3. Despite this scarcely economic traffic, the branch remained intact until the 1980s when the route was required for part of the M25 motorway.

On branch lines further west, the Slough-Windsor branch had no freight at all by the 1960s, the Windsor goods yard had been lifted and the branch itself single-tracked throughout. The Maidenhead-High Wycombe line, in contrast, formed a useful through route and carried freight, local passenger and the occasional diverted through passenger service. I remember my train being held at Loudwater on a cold snowy day in 1963, to cross a southbound coal train headed by 'Modified Hall' 4-6-0 No 6965 *Thirlstaine Hall* running tender-first.

The intermediate stations on this route were all of similar design and had active goods facilities at the time. In a curious hangover from the Beeching era the line north of Bourne End was closed and lifted. Its usefulness as a through route was thus ended, together with any potential for attracting

Above left:
Another branch whose freight traffic was healthy was that from West Drayton to Staines West, though until 1964 the freight activity had been concentrated at the northern end. '57XX' 0-6-0PT No 4697 rumbles the branch freight out of the loop at Colnbrook, having picked up empty wagons from the private sidings there. Much of the traffic was in steel reinforcing rods to a concrete works, but metal scrap, coal and books were also handled. *T. Wright*

Left:
South of Colnbrook, freight services ceased in 1955 but were restarted in 1964 with the opening of a central heating oil distribution depot on the site of Staines West goods yard. In the early years of the service the motive power was almost invariably a 'Hymek' Type 3 as seen here about to pass under the Staines By-pass bridge approaching Staines West with the incoming tanks. *T. Wright*

new passenger traffic between the residential areas at the southern end of the route and the developing 'high-tech' industrial area at the north end.

In the more rural extremes of the Western's London Division, Newbury-Welford Park on the erstwhile Lambourn Valley branch, was retained for traffic to the US Air Force stores at Welford Park, while Radley-Abingdon remained as freight-only for local coal traffic and output from the MG motor car factory at Abingdon. In 1964 my brother and I walked the trackbed from Welford Park to Lambourn, the line having already been lifted, although the halts and Lambourn station remained intact. Behind the little station at Welford Park there was a large spread of sidings, completely empty of traffic. The branch had been a last stronghold of the ex-Midland & South Western Junction Railway 2-4-0s, but by the late 1950s these had given way to GWR 0-6-0s and later North British Type 2 diesels — the only main line class light enough to be allowed over the branch.

Left:
Southern 'Q1' class 0-6-0 No 33001 has just collected a freight including two domestic coal hoppers and several grain wagons, from Scours Lane sidings, Reading and is crossing towards the SR in front of Reading Main Line West signalbox. The train has come from Moreton sidings, Didcot, a yard which was built in 1941 and disappeared completely with Beeching rationalisation. *B. K. Snow*

Above:
The short branch from Radley to Abingdon remained busy with freight for several years after Beeching withdrew the passenger service. As well as coal and some general freight there was a healthy outgoing traffic in MG sports cars until production at the Abingdon factory ceased. On 11 April 1967 North British Type 2 B-B No D6344 has arrived with the midday trip working from Oxford Hinksey Yard.
Andrew Muckley

The Abingdon branch contrasted sharply. It was traversed by a number of branch line DMU tours after it closed to passengers. Although Radley station was reduced to unstaffed halt status and largely demolished, Abingdon remained substantially intact and on various visits I always found the yard there jammed with traffic. The end of MG car manufacture seems to have been a severe blow, and piecemeal demolitions took place at Abingdon until, in the early 1980s the coal traffic ceased and the line closed. There was talk of a narrow gauge tourist line on the trackbed, but the branch was eventually lifted.

The branch from Cholsey & Moulsford to Wallingford survived in similar fashion to serve a malting plant just outside Wallingford station. The position of this plant unfortunately enabled the terminus station site to be sold off for redevelopment, so although the line passed into preservation after the malting traffic ceased, the preservation group cannot reach the original station area.

Left:
The Cleobury Mortimer & Ditton Priors Light Railway was a railway freight anachronism which survived into the mid-1960s to serve a Royal Navy Armaments Depot. Once worked by pannier tanks fitted with spark-arresting chimneys because of the risk of explosion, in the diesel era trains were worked through from Cleobury Mortimer by the RNAD's own locomotive. It is seen here passing through Stottesdon with a train of 'fruit' vans on 26 March 1965. *Andrew Muckley*

Below left:
'Parto' general merchandise vans with movable internal partitions, specially branded for operation between Watchet and the Smithfield Market, London, are shunted at Watchet by a '2251' class 0-6-0 on 14 September 1953. Before the control of wagon movements by the TOPS computer, a far greater number of wagons were controlled by manual methods. Special brandings ensured that the same vehicles were reserved for specific duties, but it might well mean that they were not intensively or cost-effectively operated. *Ian Allan Library*

Right:
A crew of at least five take a solitary wagon down the freight-only branch to Presteigne on 20 August 1964. The line's regular engine, '14XX' 0-4-2T No 1420 pulls away from Bullocks Mill level crossing with a working that even the most vehement anti-Beeching opponent would find difficult to justify. *B. J. Ashworth*

One of my favourite rural branches was undoubtedly the line from Chippenham to Calne, and I went there by rail several times during the early 1960s. Like the Staines branch with its linoleum, Calne had a major traffic source in the shape of C. & T. Harris's Wiltshire bacon and sausage factory. It received incoming pigs by rail and despatched its goods in specially branded 'Siphon' vans. By the 1960s this traffic was in decline, much of the output being moved by road.

On my visit there, freight traffic seemed healthy enough, and Calne yard was well filled with vans and open wagons. Few such branches could still boast a Saturday freight service, but Calne did, and on arrival there one chilly Saturday in 1964 I found '57XX' 0-6-0PT No 9672 waiting to depart with the freight to Chippenham. As soon as our DMU came to rest in the platform, the signal dropped and off it went. The closure of the RAF radio stations at Yatesbury and Compton Bassett dealt a severe blow to the branch traffic as did the ending of the Harris traffic. Freight services were withdrawn in the autumn of 1964 and the track in Calne yard was lifted before passenger services ceased a year later. The railway — and even the production of bacon and sausages — in Calne is now, as with linoleum in Staines, a memory.

I also travelled the Oxford-Princes Risborough cross-country route on railtours several times after regular passenger services had ended. One notable occasion was the returning AGM special from either the Talyllyn or Festiniog Railways. Diverted via Oxford because of engineering work on the direct line via High Wycombe, our return train was headed by 'Castle' No 7029 *Clun Castle* and in the early hours of Sunday morning we found ourselves wandering along the single track via Thame as overhanging trees slapped the carriage sides with wet foliage.

Some years later the centre portion of this route was removed, the big girder span over the A40 trunk road being resited on a branch in South Wales. The oil depot at Thame

was then served from Princes Risborough, while at the other end the Austin-Rover works at Cowley was served from Oxford. Both these short sections remain in freight-only use.

Of the other rambling cross-country route between Cheltenham and Banbury, all that remained was a short stretch from Kings Sutton-Adderbury at the northern end which served a feed mill. Adderbury station site was certainly an enticing spot for its buildings remained derelict for years afterwards, and it was while exploring here that I lost a gold signet ring given to me by my wife!

Further afield, Much Wenlock was another favourite location, but passenger services were withdrawn in 1962 before Beeching, and so we had to reach it by bus. Beyond Much Wenlock, the section through to Craven Arms had lost its passenger service as early as 1951, but Much Wenlock-Longville remained *in situ* for a freight service which ran as required for agricultural traffic. These workings, of course, also ended when the passenger service to Wenlock ceased, and by the time I got to know it, all the track had been lifted.

From Madeley Junction to Buildwas the line was retained for coal traffic to Ironbridge power station, together with a short divergence from Lightmoor Junction to Horsehay which was occasionally used for exceptional loads from Horsehay ironworks. At Lightmoor one day we were lucky enough to witness some freight activity in the shape of Derby-Sulzer Type 2 prototype No D5000 with a solitary bogie bolster wagon and brake van. This was on one of several mid-1960s touring holidays which I made with two friends, Keith Jaggers and Paul Chamberlain.

On reflection, in view of the infrequent nature of these freight services we had a number of lucky breaks, but none more so than when we arrived at Mitcheldean Road on the Gloucester-Hereford line one afternoon in August 1965. I had travelled this beautiful stretch of railway a couple of times before the passenger service was withdrawn in November 1964. As we drove into the station yard we were suddenly aware of the sound of an approaching train. Through the derelict and overgrown station came BR Standard '2MT' 2-6-0 No 78004, running tender-first with a short freight for Ross-on-Wye.

Here, even at the end, steam was still in charge. Our passenger trains in those final years had enjoyed haulage by 'Manor' 4-6-0s or '41XX' Prairie tanks, and with its pretty wayside stations, single track line and passing loops, it had remained truly *Western* throughout. If ever there had been a candidate for preservation, this was surely it.

The branch from Little Somerford to Malmesbury remained open for freight until the early 1960s, although passenger services had ceased in 1951. It dealt mainly with shipments of farm machinery but was interesting in that it provided a trip working for 204bhp 0-6-0 diesel shunters, several of which were based at Swindon. They also worked the Highworth branch which was retained for freight and workmen's trains after passenger services ended in 1953.

Below:
The Western Region inherited the Great Western's long-standing commitment to shunt the docks at Plymouth which dated from the time when trans-Atlantic liners had called there. Docks traffic dwindled in postwar years but the WR continued to operate the '1361' class saddle tanks which the GWR had built specially for the job. At Millbay Dock on 19 September 1957 No 1361 poses with an appropriately branded shunters' runner. *Ian Allan Library*

Indeed, this branch provided the regular sight of one of these locomotives on a passenger working.

On one of our summer tours we found a 'Hymek' Type 3 diesel-hydraulic at Cranmore with a train of tar wagons, long before the establishment of either the East Somerset Railway or the Foster Yeoman stone terminal which has reinvigorated this particular freight branch. Indeed, it was the prospect of a complete lack of freight traffic from the county of Somerset, following the Beeching closures, which prompted the development of the Foster Yeoman terminal at Merehead quarry. In this instance it was fortunate that railway infrastructure near the quarry had survived intact after Beeching. How many other such enterprises could have been established had the Beeching era clamour to close and destroy rural railways not taken place?

Right:
The goods yard entrance at Backney halt near Ballingham on the Gloucester-Hereford line speaks volumes for the decline of rural freight services by the 1960s. In August 1964 it was just months from closure.
Andrew Muckley

Below:
In the author's view the prettiest of the GWR 4-6-0s, a 'Grange', No 6860 *Aberporth Grange*, storms over Dainton Summit with a fast vacuum-braked van train. Dainton Tunnel signalbox has the distinctive oval toplights which betray its origin as a box built by signalling contractors Saxby & Farmer. A few such boxes survived on the WR, particularly on the erstwhile South Devon Railway lines, until the Beeching era. Damage resulted in the Dainton box being replaced by a modern structure. *Ian Allan Library*

CHAPTER

5 Resorts for Railfans — Reading and Oxford

Unlike most railway enthusiasts, I did not spend the 1960s going out in search of steam. As I lived alongside the Southern Region, first at Weybridge and later at Staines, I saw plenty of steam and actually started going further afield in order to find new diesels. At first it was simply days spent at the lineside near Iver station, but we quickly graduated to join the throng on the platforms at Reading General.

The easiest way to Reading from my home was by the Southern Region line via Ascot. The trains were electric multiple-units of the 2-HAL and 2-BIL types which split at Ascot, one section going to Camberley and the other to Reading. Here they terminated in the old Southern station which was adjacent to Reading General. As the Southern line entered Reading, the WR main line could be seen converging from the right on a long embankment. Once the gasworks was passed there were two double-track lines linking Southern and Western by way of inclines one much shorter and steeper than the other.

Below:
A wooden-post semaphore bracket signal on the up main platform at a most unusually deserted Reading General. In this 1950s period view the station retains its GWR gas lamp posts. *Ian Allan Library*

Atop the embankment stood Reading General, its long sprawl of dirty yellow brick buildings, dating from 1896, set behind a shambles of bus shelters and taxi ranks. Viewed from across the road, it looked more handsome, and the original Brunellian section was recognisable, topped with a handsome clock tower. Happily, the reconstruction during 1989 retained these features and improved them by stripping away some of the grime and the later alterations.

The plan shows the platform arrangement with its four main platform faces and six bays. They are shown as they were from the 1960s, when the down main platform was renumbered 4. The small parcels bay at the London end was un-numbered, so when Beeching rationalisation closed the SR station and diverted Southern EMU services into a new bay at this end, it had to take the number 4A. Traffic levels soon required further alteration and a second bay, 4B, was added.

When I first knew it the little bay siding was a parcels dock, much used by the GPO as it had direct road access. When it was altered, Post Office operations moved to an area further along Platform 4. At the west end, three bays numbered 1-3 served local trains over the Berks & Hants line to Bedwyn as well as movements off the SR line to Basingstoke, and vans stabled for loading.

The centre through road, a legacy of broad gauge double-track clearances, provided for up trains to pass nonstop, while those requiring to stop diverged on to a loop through Platform 5. The through road was signalled for running in either direction when MAS colour-light signalling was introduced in 1965. The main island platform was provided with a single bay at either end. That at the eastern end served the Reading-Paddington local service, which by the 1960s was worked by DMUs, so no run-round facilities were needed. The west end bay seemed mainly to provide a refuge for the west end station pilot.

Below:
Railways in the Reading area before Beeching (not to scale)

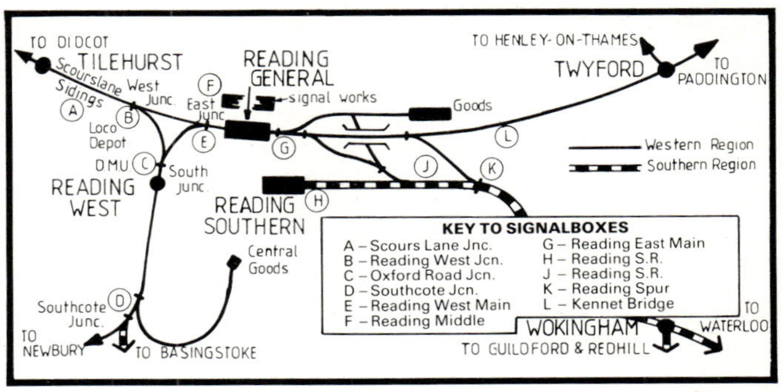

Left:
Just post-Beeching after the adoption of the corporate image the down main platform at Reading General is seen, showing the original building with its very fine clock tower, and the 1960s-built Western Tower in the background. The major change which Beeching brought at Reading was the closure of the adjacent Southern station, with its services concentrated on two bay platforms at the east end of Reading General. *Author*

Between the relief line platforms Nos 8 and 9 there was also a through road. These platforms were used mainly by semi-fast and local services, such as those to Oxford. Platform 9 also tended to be occupied for long periods by parcels trains and today still performs the same function. At its London end another bay, No 10 was also used by local services and by the parcels railcars which shuttled daily between Reading and Paddington.

Behind this platform were goods lines which enabled begrimed WR freight locomotives with indiscernible numberplates to escape the eagle eyes of spotters who always thronged the end of Platform 5.

West of the station the Berks & Hants line which provided a direct route to the West of England via Newbury, diverged to the left of the four-track main line to Swindon and Bristol. In the fork of the diverging routes was Reading motive power depot, and beyond it on the Berks & Hants line was Reading West, the third of the town's stations. It was little more than a halt, with a concrete platform of comparatively recent construction. Immediately at the platform end, the Reading West Curve diverged to form the third side of the triangle around the depot, enabling trains from the Basingstoke direction to turn west towards Oxford without reversal at Reading General. Following Beeching changes and the decision to concentrate all main line services on Reading General, of course, these trains do now go into the main station and reverse.

Below:
The platform arrangements at Reading General in the early 1960s

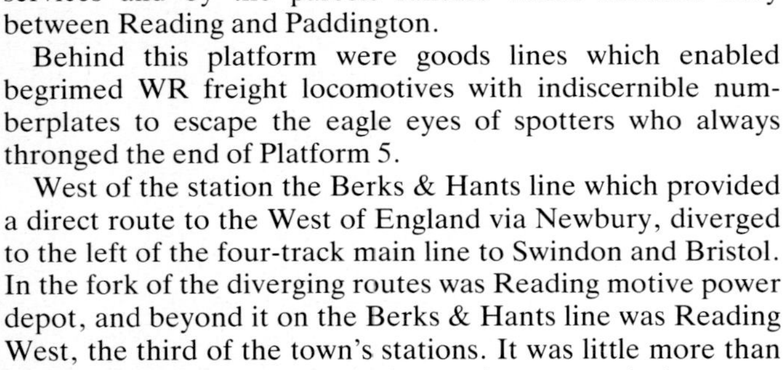

In steam days trains from the south coast to the Midlands and the North ran via the West Curve and called at Reading West while motive power was changed at Oxford rather than Reading. When the Somerset & Dorset line was closed by Beeching and several inter-regionals were transferred to run via Reading, these included the 'Pines Express'. Those of us wishing to see or travel on such trains therefore had to visit Reading West.

Reading General boasted two station pilots and in the early 1960s both of them were usually steam. At the east end there was usually a 'Manor' 4-6-0 facing towards Paddington. At the west end a 'Castle' was provided facing westwards in order that it was ready to replace any main line locomotive which might have failed or proved troublesome in the early stages of its journey down from Paddington. Reading had several 'Castles' on its allocation, but the one which I remember most clearly on station pilot duty was No 7011 *Banbury Castle* which appears to have been a Worcester-based engine at the time.

In the Beeching era, inter-regional expresses were a second-class breed. Timings were usually leisurely, there were lots of stops and protracted engine changes which added ages to the journey. Many were summer-only operations, and one such used the Reading-Tonbridge route. It came from Sandwich, Ramsgate, Margate, Folkestone and Dover, its two sections joining at Ashford, Kent. It reached Reading General just before 1.00pm and continued to Birkenhead by way of Oxford, Birmingham, and Shrewsbury. Through working to Birkenhead via the WR was, of course, a victim of Beeching's route rationalisations. I can recall this train approaching up the longer of the two inclines into Reading General from the SR, usually with a Southern Mogul or a Standard '5MT' 4-6-0 in charge. Its most distinctive feature was its southern green coaches, often Bulleid or old Maunsell stock.

The change of locomotive took place outside the station and on the day when I rode it, it rolled into Platform 8 behind Reading's own 'Castle' No 5076 *Gladiator*. It took the down relief road as far as Didcot and the 'Castle' ran well, a clear run on to the Didcot East Curve seeing the speed scarcely reduced. Didcot MPD, now the Great Western Society's preservation centre, was then a working steam shed, and it was here that on one occasion we saw the aftermath of a serious accident. Running light engine, a Stanier '8F' 2-8-0 had struck the Fawley-Bromford Bridge block oil tank train on a crossover north of the depot. The resultant fire had destroyed the locomotive and an overline footbridge. The girder spans of the bridge were quickly removed but its severed stairs stood for months as a mute reminder, while the ashen and twisted '8F' languished dumped at the end of the shed yard.

The alternative method of getting a ride behind steam in these twilight years of steam on the Western, was to make one's way round to Reading West either on foot or on the Bedwyn-bound DMU. Here the various expresses from Bournemouth to the Midlands and the North ensured a ride behind a Bulleid Pacific. On one very pleasant summer day we caught the northbound 'Pines Express' with an immaculately turned-out 'Merchant Navy' No 35023 *Holland-Afrika Line*. Returning by the southbound York-Bournemouth later in the day produced a grimy 'West Country' No 34036 *Westward Ho!* In the meantime the 'Merchant Navy' had been coaled, watered and turned on Oxford MPD, and returned with the southbound 'Pines'.

These through trains from the SR changed engines at Oxford and one had the option of continuing behind WR steam, although the onset of dieselisation saw these workings soon turned over to the most boring of diesels, the Brush Type 4s. I once took this option and went on to Leamington Spa behind No 6944 *Fledborough Hall*. It was priming badly and showered me with warm dirty water as it departed!

More interesting was a ride down the Worcester line on the 1.25pm Saturdays-only local from Oxford. Its eventual destination was Worcester Shrub Hill, calling at all stations and taking a shade under 4hr for the 60 miles! To achieve this entailed a lengthy stopover at Moreton-in-Marsh where it shunted back into a siding to allow a down express to overtake. It was usually a couple of old corridor coaches headed by a WR 4-6-0, but on dieselisation instead of being replaced by a DMU, a 'Hymek' Type 3 was provided for the two Hawksworth coaches, resulting in an extremely lively ride with breathtaking acceleration from the wayside stations.

Returning to Reading, a typical Saturday afternoon in the early 1960s would offer steam haulage on Worcester line trains although some of these were going over to haulage by 'Hymeks' and Brush Type 4s. As late as summer 1964 I recorded 'Castle' No 7025 *Sudeley Castle* on an up express, while a decrepit No 5055 *Earl of Eldon* also turned up running light engine. Streaked with rust and leaking steam, its number was painted crudely on the cabsides but its nameplates were still in place.

During 1964 resignalling work commenced at Reading General with the installation of MAS colour lights and simplification of the track layout. Before that the station had boasted some fine lower quadrant brackets. At the London end of No 4 platform there was a fine example with three dolls, two of them having the curved sighting boards to minimise the confusing background. They stood adjacent to the parcels bay and here I witnessed some amusing entertainment one Saturday afternoon.

In the days before Beeching, the railways carried livestock. By the 1960s it had apparently become uneconomic and BR wanted to shed its ageing fleet of horseboxes and cattle vans. Though some Berkshire stations such as Lambourn had enjoyed a thriving traffic in racehorses, Reading was not accustomed to equine passengers. The show began after the east end station pilot No 7808 *Cookham Manor* shoved a couple of horseboxes into the parcels bay. One of these, if I remember correctly was an ex-GWR pattern vehicle of wooden body construction. My attention was first drawn to this vehicle by the tremendous commotion which began when staff tried to extract the occupant. It sounded as if he would kick the van to pieces! Various uniformed staff members were summarily ejected from the vehicle, including one who had boasted loudly that he would demonstrate how it should be done. He left the van a good deal faster and less elegantly than he had entered it, and when I left the station they still had not unloaded the horse!

Parcels trains produced a goodly selection of 'Hall' 4-6-0s right up to the end of steam, and though I recorded several of them on film, few of my photographs are of publishable quality. A splendidly ash-coloured and leaking No 5962 *Wantage Hall* was caught waiting in Platform 8 with 1M57, an unidentified down express to the LMR. I was unable to trace the identity or destination of this train in the working timetables and it may have been a summer relief working.

The west end station pilot was No 6953 *Leighton Hall* on another occasion, and spent its time moving parcels vans from the north side of the station, across the complex trackwork to the bays on the south side. More unusual, was No 4920 *Dumbleton Hall*, complete with red-backed nameplates, which arrived on a troop special from the Basingstoke direction. Its train was of SR green-liveried Mk 1 stock.

The last day of steam at Reading depot was 4 January 1965 and I contrived to be at the station for the final Saturday of steam. It was to be a few more months before MAS came into operation, but the gantries were already complete, the signal heads bearing the white crosses which indicated that they were not yet in use. I was experimenting with black and white photography having recently upgraded to a 'new' secondhand camera. There was little steam activity in fact but I did manage a couple of shots of a well-groomed 'Tanner-one' 2-6-2T No 6161 as it paused briefly at the end of Platform 4. The Prairie was in lined green livery with copper-capped chimney and all of its plates were intact. Even the 81D Reading shedplate was clearly legible.

By this time the local shunting had passed entirely into the hands of 350hp diesel shunters and one could no longer watch a long train of wagons snaking through the subterranean passageway from Reading Southern to the WR goods shed behind a '57XX' 0-6-0PT.

Steam could still be seen occasionally at Reading for a long time afterwards. Probably my last recorded sighting, on film anyway, was on the first Saturday of the 1965 summer timetable. This was notable for major accelerations of services, particularly those to South Wales, and it also brought the usual crop of summer-only inter-regionals. While I was at Reading General No 6963 *Throwley Hall* came up through the centre road with a lengthy freight train. To see any more steam it proved necessary to go further down the line, to Pangbourne, where the inter-regionals coming round via Reading West Curve were almost exclusively in the hands of Stanier 'Black Five' 4-6-0s.

It was *Throwley Hall* and its freight train which somehow summed up the last dying blasts of steam on the Western, so intermingled as they were with the effects of Dr Beeching's wretched Report. It was particularly ironic that on this first day of vamped-up services, shortly after the 'Hall' had

Above:
In the last few days of steam at Reading '61XX' 2-6-2T No 6161 in smart green livery and with its 81D (Reading) shed plate, drifts through the down main platform at Reading General on 4 January 1965. It is framed by the MAS colour-light gantry wearing the crosses which indicate that it is not yet in use. *Author*

passed, a maroon 'Western' diesel struggled into Platform 4 having taken twice the allotted time for its journey from Paddington. It was heading a 14-coach train on one of the fast South Wales schedules. In time-honoured fashion the west end station pilot was substituted, but now it was a mere 'Hymek' Type 3 diesel instead of a 'Castle'. The diesel made a gallant departure with a very heavy train and 40min lateness to make up!

The post-Beeching era saw Reading General change comparatively little and it has remained an interesting station. As late as 1987, before platform modernisation began, I watched one of Foster Yeoman's big stone trains wend its way across the trackwork west of the station and was reminded of watching a '28XX' 2-8-0 with a train of ballast hoppers doing the same thing more than 20 years before. When I looked down, I was standing on the rough tar patch covering where the water column had stood.

Oxford

By contrast with Reading, I always found Oxford something of a paradox. Its buildings were among the most welcoming of any main line station despite being 'temporary' wooden structures which had lasted more than a hundred years. The refreshment room, a fully-fledged licensed bar with a nice 'pub' atmosphere, served a super egg and tomato roll. The station layout, with just two through platforms, and scissors-crossovers to enable trains to be double-banked at the same platform, was awkward in the extreme. North-facing bays were provided at the ends of both platforms. The bay on the down platform served local trains on the Worcester line (after Woodstock branch locals ceased using

Right:
A DMU from Bletchley has arrived in the bay at Oxford on an unrecorded date in the early 1960s. Ex-GWR vans and horseboxes are visible in the sidings on the left, looking towards Rewley Road station. This siding area became a car park after the Beeching era rebuilding of Oxford station. *Andrew Muckley*

it) and was the point of departure for my much-loved 1.25pm Saturdays-only to Moreton-in-Marsh.

A notable feature of this platform was the late survival of the GWR swan-necked gas lamps. Even after Beeching had done his worst, and the bay platform had been displaced by car-parking, these lamps continued in use, finally succumbing when the station was rebuilt. The bay platform on the up side had served Fairford branch trains, although by the time of my visits the branch had closed, and also trains from Bletchley off the old LNWR route. The former LNWR Rewley Road station area remained in use as a coal depot, rail access being by a most unusual swing bridge over the canal.

The first station out of Oxford on the LNWR line was Islip and I always treated this as an 'honorary' WR station. It had a wooden building, not unlike those on the Worcester line, but on a timber-framed platform and with an open lever frame. Perhaps its best feature was that the building was oil-lit until the end. That end came when the whole Oxford-Cambridge through route was closed from 1 January 1968. Again this was a Beeching closure and the process of reversal began in the mid-1980s. The track had been retained for occasional freight trains, and a passenger service was reintroduced between Oxford and Bicester in 1987. Traffic problems in Oxford formed part of the motive behind this move, but there has since been mounting pressure for a

Below:
This unusual notice which adorned the Didcot end of the up platform at Oxford disappeared in 1964 and BR sought information on its whereabouts. Apparently theft was suspected, and one wonders if it ever turned up. *W. L. Kidston*

Right:
Beeching's closure of Islip, first station out of Oxford on the line to Cambridge, was reversed on 13 May 1989 when a new unstaffed halt was opened on the site. The original station, seen here in July 1966, was built of timber and lit by oil lamps. The open ground frame is visible on the left beside the station building. The Oxford-Cambridge service employed the very handsome 'Derby Lightweight' diesel units, among the earliest of BR DMU designs, one of which is seen here awaiting departure. *Andrew Muckley*

CAMBRIDGE

through passenger service between Oxford and Milton Keynes via Bletchley. At the time of writing the opening of a new Islip station is just a month away.

North from Oxford on the line to Banbury, local services had also been proposed for withdrawal by Beeching. Although the stations at Kidlington, Bletchington, Fritwell & Somerton and Aynho closed in November 1964, the substitute bus service could not cope with the other intermediate stops. So Tackley halt, Heyford and Kings Sutton had to remain open. They were swiftly reduced to unstaffed halts and Kings Sutton passed into the LMR as part of boundary alterations which redrew the WR frontier just north of Aynho station.

Kings Sutton was a super place to spend an afternoon in the mid-1960s, with a succession of inter-regional trains, summer-only expresses and steam-hauled freights passing its little platforms. Its buildings were a much more ornate version of those on the nearby cross-country line to Kingham. There were three such stations – Adderbury, Bloxham and Hook Norton – plus a larger version at Chipping Norton. The passenger service between Kings Sutton and Chipping Norton had ceased in 1951 after the line was damaged by a storm, but Chipping Norton continued to be served from Kingham for another 10 years. At the northern end the track was retained as far as Adderbury for occasional trains of farm feeds and fertilisers.

On the down platform at Kings Sutton there was only a small brick-built waiting shelter, and happily this still survives, although the main station building was demolished and replaced by a glass shelter. I spent an afternoon there before the demolitions began, although signals and station lamps had been removed and lay scattered about. In the warm sunshine lizards scampered about on a pile of old rail fixings and the air was heavy with country smells. Making a loud clanking noise, a 'WD' 2-8-0 rattled through the station heading westwards, light engine. It was the last time I saw one of these locomotives in traffic.

At Aynho station and Kidlington the main station buildings survive in private use, although in poor condition and with their platforms removed. The similar building at Heyford was luckier and survived until the mid-1980s when it was acquired for preservation at Didcot Railway Centre. It has been removed stone by stone and will doubtless rise again in due course.

Two Oxford branches, Kidlington-Woodstock and Oxford-Fairford had closed even before Beeching while the closure process for Radley-Abingdon had been set in motion before the Report was published and closure took place shortly afterwards, in September 1963. Although the Abingdon passenger service had been worked by single-unit diesel railcars the other two branches had closed before being dieselised.

The Abingdon line remained in use for freight (coal and MG cars), as did the Oxford-Princes Risborough line as far as Cowley, the British Leyland car works. The Fairford branch

Above:
In the early 1960s the Worcester line seemed to always be basking in sunshine, with pristine stations and smart trains. This view of Kingham is typical with an immaculate 'Castle' No 7031 *Cromwell's Castle* arriving on an up express. The spotlessness of the platforms is an object lesson. *Ian Allan Library*

Above right:
Handborough for Blenheim was the first of the typical Oxford Worcester & Wolverhampton Railway wooden stations out of Oxford. Like the others, it lost its little canopy in the 1950s but seemed to escape repainting. When it was used as the terminus for the funeral train of Sir Winston Churchill, its faded and peeling GWR paint was an embarrassment and the whole building was obscured with purple and white drapes. This is the view towards Kingham on 11 April 1967, after closure and 18 months after the funeral. *Andrew Muckley*

Right:
Another original OW&W building on the up platform at Adlestrop, on the day it closed, New Year's Day 1966. The GWR had made improvements here, providing a stationmaster's house (left), pagoda shelter on the down side, and longer platforms. The signalbox and goods shed at the far end of the station had gone by this time. Note the absence of platform lighting. *Andrew Muckley*

HANDBOROUGH
FOR BLENHEIM

OLESTROP

CH ING CAMPDEN
LONDON

was retained as far as Witney for freight, and years after Beeching I had the opportunity to travel all these lines on various railtours and 'last trains'. Only the line to Cowley is still in use.

So, although the effects of Beeching were severe in this part of Oxfordshire, with local services having to be retained on both the Worcester and Banbury lines, and with the Bicester line being resurrected, one cannot help but feel that the closures which did take place did not achieve very much. After all, destaffed stations can surely contribute to revenue without a significant adverse effect on costs or journey times.

Left:
At Chipping Campden the main building was on the down platform, but here the original OW&W waiting shelter on the up platform also survived in all its decorated splendour. As at Handborough the platform lighting was by Tilley oil lamps until the end. The level crossing is now lifting barriers and nothing else remains at this site. *Andrew Muckley*

Below left:
The view towards Oxford at Chipping Campden as 'Castle' No 7013 *Bristol Castle* takes a run at the bank with a down express on 23 May 1962. All the neat detail of the waiting shelter is clearly seen. *Ian Allan Library*

Below:
Chipping Campden had a spacious goods yard with a wooden goods shed dating from broad gauge times. Standard '2MT' Mogul No 78009 hustles a down freight through the station in 1962. Since closure even the factory building in the background has vanished. *Ian Allan Library*

Bottom:
This vast expanse of deserted platforms was at Honeybourne, an operationally important station with little local traffic. Following Beeching its fortunes changed and it became one of the first Beeching closures to be reversed. All the buildings have gone, and the down platform (right) has been reinstated to serve what is now a single track line. *Andrew Muckley*

CHAPTER 6 A Dieselised Western

Dieselisation was not a feature of the Beeching Report, as it had been set in motion by the 1955 Modernisation Plan and was well under way by 1963. However, since it took place concurrently with the Beeching closures, and since the elimination of steam and the consequent cascading of diesel units as a result of branch closures were planks of the Beeching strategy it is certainly relevant to devote a chapter to the subject. Most importantly, from the viewpoint of this book, the transitional period, with steam and diesel side by side was an important part of the WR's attraction in the early 1960s.

The Western was quite late in commencing its dieselisation. Though other regions had started to receive DMUs as early as 1954/5, the WR had only a handful of elderly GWR railcars. It was not until wholesale deliveries of 'suburban' DMUs in 1957/8 that complete areas could go over to diesel operation. In general terms, units built by Birmingham Railway Carriage & Wagon Co (BRCW) (later Class 118) went to the West of England, those by BR Derby (later Class 116) went to South Wales, and those by Pressed Steel (Class 117) came to the London area. Cross-country units with superior accommodation and miniature buffet facilities were built by BR Swindon and by Gloucester Railway Carriage & Wagon Co and were distributed throughout the Region, while both Gloucester and Pressed Steel built suburban-pattern single-unit railcars for branch line services. The majority of the units allocated to the Western Region were compatible for multiple-unit operation, driven from the leading cab, and accordingly they quickly became mixed up, particularly in areas such as around Bristol were the duties of types from all three areas tended to overlap.

The main difference between the earlier units built by Gloucester and Derby was that they carried small destination screens in the roof dome, and a two-character headcode display below the centre windscreen. Even before they entered traffic the four-character headcode had been adopted as standard, and designs such as the units from BRCW and Pressed Steel incorporated the four-character panel in the roof dome and had the destination box added, almost as an afterthought, in the centre windscreen. The four-character headcode consisted of three elements; a number indicating the class of train, a letter indicating the area (A for London, B for Bristol etc), and a two-digit number specific to the route. Thus, 2A82 was a branch passenger train in the London area, specifically West Drayton-Staines West.

When the Gloucester single-unit cars were delivered in 1958 the majority came to the London area to replace steam-hauled auto-trains and GWR diesel railcars on the

Right:
The driver's-eye view. With the advent of DMUs, passengers were provided with a unique forward view as demonstrated by this unit working the Kingsbridge branch in South Devon. The railcar is coming to a halt at Gara Bridge station and the driver's right hand can be seen on the brake handle. *Ian Allan Library*

Left:
A close-up of the delightful Gara Bridge signalbox and level crossing seen in the previous view. All the ingredients of a classic GWR country station survived here right into the diesel era. *D. B. Hart*

18000

Thames Valley branch lines. The GWR cars were then concentrated on Worcester, where they worked Severn Valley and Tenbury Wells area services. With the arrival of the Pressed Steel single-units in 1960, these cars took over the London area duties, ousting the Gloucester cars to the Midlands, Bristol and the West Country.

Both the Gloucester and Pressed Steel singles had been provided with small quantities of matching driving trailer cars, unpowered and with a driving cab at one end, to enable the operation of two-car units. These were thought desirable to cope with peaks of traffic, since the WR had no two-car units in its fleet. In practice few branch services required the second car, even at peak times, and if it was required then it would run in traffic all day to avoid the complication of coupling and uncoupling the vehicle. The fact that the two-car units so formed could not be gangway-connected made the issuing of tickets by the guard much more difficult. The driving trailers were used on the Staines branch for a year or two before it was decided that a single car was adequate. They survived rather longer on the Henley and Marlow branches and continue in occasional use on the Windsor branch weekday service. The latter, rather special case, actually justifies a three-car unit for the Sunday service!

At the 'top' end of the DMU fleet, initially, were the two types of cross-country three-car units built by Gloucester and Swindon. These were near identical in appearance except for the cab ends, the Gloucester units having that organisation's standard arrangement with two-character headcode. The Swindon units had the distinctive and handsome end styling used by Swindon on earlier units built for service in Scotland. This design used neither two nor four-character headcodes but had instead electric headlamps which were placed to represent the positions used on the old steam locomotive headlamp code. Later-built examples carried the four-character display uneasily on the lower front panels.

In both types seating was in open saloons reached through transverse vestibules, with gangway connections provided only within the set, so passage between two units running together was not possible. The centre car was an open trailer second equipped with a four-seat miniature buffet at one end. Both power cars were equipped with two BUT/AEC 150bhp engines, one car having a guard's compartment, and the other a first-class saloon in addition to five second-class seating bays. Seats had head and arm rests, and folding tables enabled passengers to take light refreshments at their seats.

Both types were painted in the new DMU livery of bright green, with cream bands at waist and cantrail, and a cream 'V' embellishment on the ends, following experimental finishing of some cars to assess reaction.

The Swindon units first ran in public service on 10 March 1958 between South Wales and Birmingham, on workings previously operated by steam or by the six-car Inter-city units built at Swindon for the Scottish Region. The first steam services to be replaced were the 8.50am Cardiff-Bristol and 10.22am return, and the 4.47pm and 8.15pm Cardiff-Birmingham (Snow Hill) and the 5.40pm return. The Inter-city units were displaced from the 12.10pm and 8.15pm

Left:
Motive power policy on the 'Big Four' during the 1940s differed considerably and World War 2 prevented any significant developments in the first half of the decade. While the LNER considered electrification, and the Southern and LMS experimented with diesels, the Great Western ordered a gas turbine locomotive from Brown Boveri in Switzerland. It was delivered after Nationalisation and took the BR No 18000. Though it proved powerful and economical over steeply-graded routes, it was 'thirsty' on the Western's flatter stretches and suffered numerous failures and protracted periods under repair. It was quietly disposed of at the time the new 'Warship' diesels were entering traffic and is preserved in Vienna. *G. J. Jefferson*

Below left:
Among the earliest DMUs to be allocated to the Westen Region were Derby-built three-car suburban sets sent initially to South Wales for service on the numerous branches and local services. Seen in March 1966 this unit has acquired the later dark green livery and yellow warning panels. It is about to call at Foley Park halt near Kidderminster on the Kidderminster-Bewdley service. This was another victim of the 'post-Beeching' closure series in 1970. *Andrew Muckley*

Below:
Among the first new DMUs in the London area were Gloucester RC&W single-unit railcars which displaced the aged GWR diesel cars for further service at Worcester. Here, a Gloucester railcar is leading its matching driving trailer car into Staines West station as the signalman waits to collect the single line staff. The cars wear the early light 'electric green' livery with cream lining and white roof ends.

Cardiff-Birmingham (Snow Hill) and the 12.25pm and 8.30pm return.

As deliveries increased operations were extended to include some semi-fast workings into the London area, such as a mid-afternoon Oxford-Paddington trip which worked back during the evening rush hour. For this and similar heavily used workings, the units lacked sufficient accommodation. Indeed, on occasions the units had been known to trail a single main line coach when demand had warranted. Three of the Gloucester units were therefore strengthened by the addition of an ex-GWR Hawksworth composite converted to run within the DMU sets. This incongruous addition certainly did not enhance the modern image of the units, and even though painted in matching colour scheme its presence could not be disguised. All three Hawksworth coach conversions were consigned to the Swindon dump in the late 1960s.

It seems to have been the fate of DMUs on intermediate distance services to bring about their own downfall. For just as the Cardiff-Birmingham trains worked by the original

Below:
For comparison, a scene at the same terminus little more than a year later. In 1960 the Gloucester cars were replaced by similar Pressed Steel vehicles. The main differences were the four-character headcode panel and the newer dark green livery. Driving trailer car No W56286 is nearest the camera at an immaculate Staines West which had just won the Western Region's Best Kept Station Award. *Dr K. A. Jaggers*

Right:
Not all went smoothly with dieselisation, but the situation illustrated here was a little more complicated than it appears. A three-car cross-country unit failed at Newton Abbot while on a working to Torquay on 27 March 1962. The unit had to be taken out of traffic and Newton Abbot provided '41XX' 2-6-2T No 5153 and a Collett 'B' set of coaches, supplemented by a Gloucester DMU driving trailer surplus from the Kingsbridge branch. The three illustrations show this unlikely ensemble at Newton Abbot and on the return journey to Torquay.
Ian Allan Library

5153

6
CAR
515

GWR diesel railcars, had succumbed to locomotive haulage in the 1930s, so the cross-country DMUs were ousted from their duties on Cardiff-Portsmouth services. Here the cross country units shared the Bristol-Portsmouth section with the Southern Region three-car Hampshire DMUs while the newer Swindon-built Inter-city units were favoured for through Cardiff workings. The Inter-city units were among the last conventional DMU designs produced, and incorporated a gangwayed glassfibre cab end with wrap-around windscreens similar to that on the ER Clacton electric units.

Although designated as Inter-city, these units were generally to be found supplementing or replacing cross-country units on longer journeys. In the final months of the Minehead branch, one such unit was employed on a Saturday through working to Paddington. Almost from the beginning, cross-country DMUs were to be found on rural branch and local services, and over the years little use was made of the miniature buffet facility. Equally, weekend excursions often employed surplus suburban units not suited to long non-stop runs, and with makeshift snack facilities. Yet the buffet sections were not removed until the early 1980s, and then only on the Gloucester units, which by then had found a new lease of life on Reading-Gatwick services. The buffet area was then rebuilt as a rack for airport passengers' luggage.

Far from increasing the capacity of the units, a number had their trailer cars removed, reducing them to two cars and so increasing the power available for heavily graded routes. In this guise, and fitted with an additional headlight, some of the Swindon units were employed on Central Wales line trains.

Remarshalling of cross-country units with 'foreign' trailer or motor cars often produced untidy rakes of assorted cars on WR services, reminiscent of the multitude of coach profiles which once featured on some GWR expresses. Some such rakes were no doubt brought about by failures, such as one which occurred at Barnstaple Junction when the rear coach of a six-car rake bound for Exeter failed. The motor brake second of a Swindon cross-country unit was immobilised at the platform, apparently with seized brakes. It was eventually uncoupled and left at the platform while its partners departed for Exeter, the gangway connection of the trailer car flapping in the breeze!

In the aftermath of Beeching, large tracts of ex-GWR territory in the Birmingham and Wolverhampton area passed to the LMR. The DMU allocation at Tyseley was included and in due course a rationalisation of DMU types saw the Gloucester vehicles concentrated on the Western Region, while the Swindon cross-country sets went first to the East Midlands and subsequently to Scotland. Though re-engined and refurbished in the mid-1970s in order to keep them in traffic until 1990, the Swindon units were victims of the purge of asbestos-insulated vehicles. The last examples were withdrawn in 1989, by which time withdrawals were making heavy inroads into the Gloucester RC&W sets.

Very much at the bottom end of the DMU fleet were the four-wheel diesel railbuses based on German practice. They were acquired as an experiment to further reduce operating costs on lightly-used lines. BR bought 22 of these lightweight vehicles from five different manufacturers. Four went to the Western Region and these appeared in 1958 from the factory of AC Cars Ltd at Thames Ditton. This manufacturer was best known for specialised production of sports cars and was a newcomer to the railway field. The AC railbus was 37ft long, weighed 11 tons and had second-class seating for 46 passengers. The four-wheel underframe of 19ft rigid wheelbase was powered by a single BUT 150bhp underfloor engine of the type used in pairs on many of the large bogie railcars. The styling of the long overhanging body was light and modern with large windows and a single air-operated sliding door in the centre of each side. These doors constituted the only entrance to the car, the driver's cab being reached from the passenger saloon. Seats in the two saloons faced outwards from the central entrance which contained two large luggage or standee areas. The cab occupied the left-hand corner of each end, and alongside it was a curved seat for three passengers which faced inwards. Seating and interior decor was very much on road vehicle lines. The sides of the car extended down over the solebars, and only rudimentary strip-metal buffing gear and a link-and-pin emergency coupler were provided. Unlike German railbuses, the ability to run in multiple or haul trailers was not part of the BR specification.

The four WR vehicles were put to work on the Tetbury and Cirencester services from Kemble. These inaugurated new services in February 1959 amid publicity which even provoked TV news coverage. A small combined poster and

Below:
The handbill which announced introduction of railbuses on the Kemble-Tetbury and Cirencester branches. Another halt was subsequently opened at Park Leaze on the Cirencester branch. *Author*

timetable with an attractive drawing of a speeding railbus was among the items prepared by the WR to announce the changeover from auto-trains. Passengers who had been used to the mixed trains which ran to Tetbury until very late and which took a long break for shunting at Culkerton must have found the railbuses quite revolutionary! The services employed three of the vehicles, two of them being housed overnight in the former branch locomotive sheds. Each morning the Tetbury branch car would transfer to the Cirencester branch, while the railbus from that line ran empty to Swindon for servicing. The third car would then run down from Swindon to take over the Tetbury service. The fourth vehicle remained at Swindon as a spare. All four cars were finished in the early railcar livery of 'electric green' with a cream bow-wave on the ends. The green was carried up over the roof domes and a polished metal strip replaced the usual waist lining.

Although the railbuses do not appear to have suffered from any worse mechanical problems than their larger sisters, they created their own operational problems, while advantages such as the use of conductor-guards were equally practicable with conventional DMU stock. The railbuses could not be relied on to operate track circuits due to their low weight, and so they were only able to run empty on main lines. Lack of conventional couplers and buffers on all but the German vehicles meant that the cars could only be used singly and this restricted them to routes with minimal fluctuations in traffic. On occasions such as market days when extra shoppers came to town, the Cirencester railbus could be 100% overloaded.

Yet in other respects they were ideal. A passenger on the Tetbury train could not fail to notice that the same close association between local passengers and crew existed on the railbus just as it had done in the leisurely days of the auto-train. For its railbus services the WR opened new rail level halts at Churchs Hill, Troublehouse, and Chesterton Lane, following later with another new stop at Park Leaze. The new halts were small timber platforms and represented the only significant attempt to make use of the railbus' folding steps and the 'bus stop' operation idea. The ramshackle timber station at Culkerton was also reopened although the new Churchs Hill halt was nearer to Culkerton village. In practice the railbuses made only occasional stops at the old station, which remained locked and boarded up.

The railbuses acquitted themselves well and succeeded in bringing about increased receipts and reduced costs. However, with few exceptions, the routes to which they were assigned were already lost causes. Nothing short of a miracle

Below:
The Cirencester railbus waits at Kemble and wears the wrong destination. Note the laden trolleys showing a healthy level of sundries traffic. *Andrew Muckley*

Above:
The North British Type 2 diesel-hydraulics performed a variety of duties in the West Country, including acting as pilots over the South Devon banks. No D6331 wears express passenger headcode as it assists a 'King' 4-6-0 over Dainton Summit. *Ian Allan Library*

could have turned losses into profits where there just were not sufficient potential passengers. Alas, miracles were not to be found in the early 1960s and after less than four years of railbus operation, publication of the Beeching Report heralded the end of the experiment. Closure notices were posted and 6 April 1964 brought the official closure of the branches from Kemble. The Cirencester branch might have survived longer had it not been so conveniently partnered with Tetbury. However, with the Gloucester main line reduced to 'branch' status, and Kemble itself barely surviving as a local railhead, there would have been little future for it. Cirencester residents must now find it more convenient to drive into Swindon and join an IC125 service.

The Western Region vehicles were stored at Swindon for a while, before two were despatched to Yeovil to run the Town-Pen Mill shuttle service. The other two went to Cornwall where they worked a similar duty between Bodmin North and a newly built exchange platform at Boscarne Junction on the former LSWR Wadebridge line. Boscarne platform was closed on 18 April 1966, but reopened a month later, only to close again on 30 January 1967 when the Bodmin North service was withdrawn, and the short-lived Yeovil service succumbed in November 1965, all as a result of Beeching.

I have dealt with the DMUs at some length, both because their use and deployment was inextricably linked with branch lines and local services affected by the Beeching proposals, and also because their history on the WR is less well documented than that of the main line locomotives.

To turn now to the main line locomotive fleet, there is no need to re-tell here the background to the Western Region's adoption of diesel-hydraulic traction in preference to the diesel-electrics used elsewhere. As a result of this policy decision the WR had, by the early 1960s, a unique, distinctive and charismatic diesel locomotive fleet. The backbone of this fleet was formed by 70 'Warship' class Type 4 diesels (Type 4 denoted a locomotive of the power range 2,000-2,750hp) for express passenger and freight work. One of the advantages of diesel-hydraulic traction was superior tractive effort which suited the locomotives to freight use as well as express passenger trains. Their comparatively light weight was not seen as a disadvantage in stopping unfitted freight trains since the unfitted freight was in decline in favour of full or part vacuum braking. The lack of space for the addition of train air-braking was instrumental in their early demise.

The 'Warships', successors to an abortive 12-wheel Type 4 design of which only five were built by North British, were supplemented by a fleet of distinctive pug-nosed Type 2s (1,000-1,500hp) built by North British. These, often known as 'Baby Warships', numbered 58 examples and were initially allocated to the West of England to enable complete elimination of steam in Cornwall and then gradually eastwards. They superseded steam on many local services and branches in Cornwall, working both passenger and

Above right:
North British Type 2s also found use on local passenger services before sufficient DMUs became available. The 1Z48 headcode gives the clue that this working at Bishops Nympton & Molland on the Taunton-Barnstaple line is not a normal local train. No D6336 has charge of a special conveying the Beatles in connection with the filming of their Magical Mystery Tour movie. *J. Spencer-Gilks*

Right:
Eventually the Type 2s settled to work primarily on china clay workings over various Cornish branches, of which that to Fowey was by far the most heavily used. In September 1964 one of the class slows to exchange single line tokens at Fowey having come in from the St Blazey direction over a section of line reserved for freight only. In a bizarre Beeching era move, this section of track was actually lifted and converted into a private road so that lorries could bring in the china clay.
Andrew Muckley

freight services. They also became well known as pilots for steam and diesel-hauled trains over the South Devon banks.

In Cornwall they appeared on local passenger services on the main lines and on branch services to places such as Fowey and Newquay. A pair would also work the summer Saturdays-only, through coaches of the 'Cornish Riviera' to St Ives in the short period between the end of steam and the summer of 1966 when run-round facilities at St Ives were removed. It will be recalled that this was one of the branches proposed for closure by Beeching but subsequently reprieved by the Minister of Transport. Rationalisation here was then so thorough that not only would there never again be the facility to work locomotives through to St Ives, but the platform length was reduced such that it became inadequate for the length of trains required to operate the peak summer traffic.

As increasing numbers of DMUs became available, the Type 2s were concentrated on freight traffic, particularly within Cornwall where they became synonymous with china clay traffic, milk trains and such local and branch freight services as survived the initial impact of Beeching. Inevitably, a number of locomotives became surplus in the West Country and they gradually spread their sphere of influence eastwards, with allocations to Bristol and later to Old Oak Common. The latter allocation was mainly in order to replace the pannier tanks on empty carriage workings to and from Paddington, but they also became familiar from about 1965/6 on local freight workings.

One such duty was the oil tank train from Ripple Lane to Staines West. This comprised about 14 four-wheel tank wagons containing central heating fuel and originated at Thameshaven on the Eastern Region. It was worked from Ripple Lane to Acton Yard by Eastern Region (ER) motive power, and from there to Staines West by a WR locomotive. On at least one occasion, the ER locomotive worked throughout and it was an English Electric Type 4 (later Class 40). The branch was 'red' category which meant it could take the heaviest diesels, but despite this there was only one recorded instance of a 'Warship' and one of a 'Western' on the oil train. Usually it was a 'Hymek' or a NB Type 2. A fair amount of shunting was required at Staines West in order to remove empty tanks from the single siding, stable the loaded tanks, remarshall the two barrier wagons and ready the empties for departure. There might also be too many loaded wagons to fit the siding, in which case four or five would be unloaded immediately and added to the returning empties. While all this went on the locomotive would spend several hours at the terminus, often arriving soon after 8.00am and not departing until after lunchtime.

Mention of the 'Hymeks' brings me to the Western Region's diesel-hydraulic Type 3 locomotive. A large class of 101 examples, built by Beyer-Peacock in Manchester and introduced in 1961, they were truly mixed traffic locomotives rated at 1,700hp. They were equally at home on freight, express or even local passenger workings and quickly made a big impact on the WR scene, though they were less common in the West Country, particularly beyond Saltash Bridge.

Their first regular duties were on South Wales-Paddington services where they ousted 'Kings' and 'Castles' from even the prestige expresses, and despite their modest power rating

Above:
There were isolated instances where pannier tank locomotives were replaced by 204bhp 0-6-0 diesel shunters for shunting and local trip workings. Very occasionally these locomotives were required to work a passenger duty, and one such regular turn was the Swindon-Highworth workmen's train. Public passenger services had ceased in March 1953 but freight and workmen's trains continued into the Beeching era. Here, No D2195 is approaching Hannington with the two-coach train on 18 June 1962. *D. H. Ballantyne*

were to be found on fast expresses loaded to as many as 14 coaches. Only when the 'Western' class entered service a year later did the 'Hymeks' see some relief from these gruelling duties. They then took over the express passenger services on the Paddington-Worcester route.

The 'Hymeks' became universal in the early 1960s, appearing on freight, parcels, and even two-coach passenger turns. They replaced the WR's large fleet of 'intermediate' steam power, primarily the 'Hall' 4-6-0s. They turned up everywhere, and as spotters and enthusiasts, how we hated them! Yet, on reflection I had more exciting journeys behind 'Hymeks' than I ever did with 'Warships' or 'Westerns'.

Soon after the 'Warships' entered traffic the WR had identified a need for a more powerful diesel-hydraulic locomotive to enable further acceleration of the express services. There were technical problems in uprating the 'Warship' design, so Swindon embarked on a development of the design, installing the increased power on a 12-wheel chassis. Great attention was paid to the body styling in order to shed the 'German' influence of the 'Warships' without adopting either a box shape or the 'bonneted' styling which had found favour elsewhere on the diesel-electrics. The result was the first of 73 'Western' class locomotives, introduced towards the end of 1961.

The 'Westerns' were long and stylish, their body profile cleverly matched to that of the BR Mk 1 coaches with which they were to work. There was another striking feature, too, namely the initial experimentation with colours. I well remember my brother returning from a trip to West Drayton and reporting that he had seen a new locomotive, No D1000 *Western Enterprise*, and that it was still in undercoat. It was, in fact, experimentally finished in a pale fawn colour called 'desert sand'. Later examples appeared in maroon and in green, before the maroon livery, which matched the standardised coaches, was finally chosen. With No D1015 *Western Champion* the Western had one further experiment,

Left:
One of the first batch of 'Warship' Type 4 diesel-hydraulics built without four-character headcode, No D812 *Royal Naval Reserve 1859-1959* heads a freight down the relief line at Slough. Note that the centre bay track is being removed — the platform was subsequently extended. For a time during the mid-1960s Ford trucks built at nearby Langley were loaded at Slough station. *P. J. Lynch*

The face of the dieselised Western at Laira Depot, Plymouth, in April 1962. A D800-series 'Warship' with four-character headcode panel stands alongside the first 'Western' No D1000 *Western Enterprise* in its unique desert sand livery. *Brian Haresnape*

Above:
On delivery to the Western Region from Beyer Peacock, Manchester, brand-new 'Hymek' Type 3s Nos D7035/6 pause in the centre road at Oxford on 30 May 1962. Before long it would be their regular stamping ground on services to Hereford and Worcester. *Ian Allan Library*

painting it a rich caramel colour which was apparently Stroudley's 'improved engine green' as used by the London Brighton & South Coast Railway many years previously. This distinctive machine was always my personal favourite, but I have yet to see a colour transparency which captures the true shade.

In their early years the 'Westerns' earned their keep on the principal expresses from Paddington to South Wales, Birmingham and the West Midlands, as well as to a lesser degree on the West of England route. Apart from their superior power, which suited them to heavy freights, such as the aggregate trains which became a feature of the WR from the late 1960s, the 'Westerns' had one other advantage. Their large body had space for installation of train air braking, which the 'Warships' did not. This, as much as the decision to eliminate diesel-hydraulic types, sealed the fate of the 'Warships', although the 'Westerns' were not to outlast them by very many years.

Little more than 10 years after the WR took the decision in favour of diesel-hydraulic traction, came the announcement that diesel-electric locomotives would become standard throughout BR. So, the Western's distinctive traction was doomed and this decision, too, was inextricably entangled with the effects of Beeching. Rationalisation elsewhere, and electrification of the West Coast main line through to Glasgow would mean that diesel-electric locomotives would become surplus to requirements. These could be transferred to the WR to displace the hydraulics. The actual implementation of this decision falls outside the scope of this book, but the signs of this change of direction were evident as soon as Brush Type 4s (later Class 47s) started to work on to the WR from other Regions.

Yet, the final fling with diesel-hydraulics, and probably the most expensive and wasteful of its type, was taking place at Swindon even as the decision to abandon hydraulics was being made. At the lower end of the motive power scale there was a perceived need for a light 'road switcher', to use American terminology. This was to be a more powerful shunter than the currently available 204bhp and 350bhp types and with a road speed sufficiently high that it could be employed on trip freight workings over main lines.

The result was the Swindon-designed Type 1 0-6-0, of which 55 examples were built and the first appeared in 1964. Most went initially to Bristol and South Wales, but even as they entered traffic the Beeching closures were systematically wiping out the very services for which they had been built. A few found their way up to the London area and they were certainly interesting-looking locomotives, with their centre cab and full-depth yellow buffer beams. Even more strangely, a motive power policy plan of the time, which laid out certain standard types of locomotive in each power category actually saw these as the standard Type 1 design. As such, they would have spread diesel-hydraulic traction off the WR at precisely the time when it was being eliminated on its home territory!

In fact, the Type 1s were overtaken by events, the WR having less and less work for them. To us, perhaps, their attraction lay in their association with a particular type of traffic — branch and local freights — despite the fact that they had done comparatively little of that work. Within four years of their entering service the WR wanted rid of them and a number were transferred to Hull Dairycoates where it was felt that freight operations in the Northeast might find a use for them. Most were soon sold out of service to coal mines and steel works where they served their new masters well for some 15 more years. A number have since passed into preservation, where they now find employment on works trains, demonstration freights, and even occasional passenger trains, on the type of rural lines for which they were built.

To sum up, the combination of Beeching and dieselisation changed the face of the Western Region. Those branches and rural routes which Beeching either did not want to close, or was not allowed to close, were able to be reduced to 'basic railway' status because the DMU required no run-round facilities or complex trackwork, because the elimination of freight services meant that trackwork could be further reduced to the point where signalling was not needed either, and because sale of tickets on board the DMU enabled the elimination of station staff and the consequent destruction of station buildings.

Top:
Not unusual in the early years of dieselisation was the need for an occasional helping hand. A failed 'Western' diesel in standard maroon livery is hauled past Westbourne Park, the last few yards into Paddington by a travel-stained 'Hall' class 4-6-0, in November 1964. The notice that 'Enginemen must not let out water on this bridge' might have been more carefully worded! *Andrew Muckley*

Centre right:
Last and shortest-lived of the Western's diesel-hydraulic designs were the Paxman-engined 650bhp Swindon 0-6-0s. Brutish and powerful machines intended for heavy shunting and trip freight work, Beeching disposed of most of their suitable duties just as they were introduced. For a while they found favour on such surviving services as the local freight turn from Gloucester down the Nailsworth branch. No D9527 is seen at the Nailsworth terminus on 3 January 1966 alongside a yard crane from another era. *Andrew Muckley*

Right:
The Swindon Type 1s were initially concentrated on Bristol and in South Wales, where No D9541 was photographed passing through Platform 7 at Cardiff General on 10 May 1966 with a freight from Cardiff Newtown to Llantrisant. Many of the class were sold to the National Coal Board and British Steel where they gave some 15 years' service before passing into preservation. *R. E. Masterman*

CHAPTER 7

Bad Day at Black Dog

As a teenager, my sister was a keen racing cyclist and one of the regular events which she entered was known as the 'Bath Road Hundred'. It began on the main A4 road at Hungerford and the course lay westwards from there. This, and other family reasons caused us to make a number of weekend visits to Wiltshire and Gloucestershire in the early 1960s.

It was during one of these journeys down the A4 that I first noticed a curious little station situated where a single track branch crossed the main road just outside Calne. Investigation of maps and the railway timetable soon revealed this to be Black Dog halt on the branch line from Chippenham to Calne. On the next visit we paused briefly to look at the place and the view of its neat little cream and chocolate building, glimpsed through the trees, is one of my favourite memories of the Western before Beeching.

My brother and I spent much time trying to imagine what sort of trains might work through this bucolic place. We decided it was the perfect location for GWR diesel railcars, and even wrote to BR to enquire. When the reply suggested that the passenger trains were 'diesels', and that there were still some steam trains, then a visit by rail became essential. I think Roger must by then have been engrossed in Saturday football matches, because I made nearly all my early visits to Calne alone.

Below:
In its final years Black Dog halt epitomised the rural branch line scene. During my several visits in the early 1960s I was the only passenger to board or alight there. The people who built it no longer needed it, and there were few houses nearby to provide any traffic. Yet this halt had been so very different from the usual. Built as a private station with its own goods siding, it had served a branch line whose traffic in both passengers and freight had far exceeded most 5-mile rural branch lines. Yet under Beeching its fate was the same. In this view towards Calne, the halt, with the 'shed' built for Lord Lansdowne, is on the extreme right, a path behind it having once provided a direct route to Bowood House. To the left, the siding is still in place, but empty, and beyond it is the handsome stone cottage built for the stationmaster by the Lansdowne estate. In 1950 it was sold to the resident railwayman, Leading Porter Lovelock, and he in turn sold the outbuildings which were converted to form the bungalow on the extreme left. *D. Lovelock*

The first of these revealed that we had, of course, been entirely wrong in our guess regarding the motive power. While the GWR railcars might have looked 'right' amid the Calne branch surroundings, it is doubtful if they ever worked over the branch — certainly no one to whom I have spoken can recall even a single such incident. By the time of our visit the few survivors were all based at Worcester. The Calne branch was in fact being worked by a variety of modern three-car diesel multiple-units based in Bristol. On this particular Saturday afternoon it was a Gloucester cross-country set.

I rode through to Calne and took a few photographs there, including the branch goods departing behind a '57XX' 0-6-0PT. It seemed a rather pleasant terminus, with a large, interesting goods yard and private sidings into the factory of C. & T. Harris, manufacturer of pork pies etc. This much I could discern from the name emblazoned vertically on the tall brick factory chimney. The station itself had a neat, standard GWR red brick building, but its appearance was spoilt by a large extension built of unpainted corrugated iron. This was adorned with notices addressed to RAF personnel bound for Yatesbury and Compton Bassett RAF stations.

Curiously, the station building also seemed to be given over largely to RAF use, and the booking office and waiting room were housed in a separate wooden building, reminiscent of the small stations on the Oxford-Worcester line. The goods sidings were full of wagons and vans, even after the pannier tank had departed with its sizeable load, and the three-car DMU was almost lost at the end of the long passenger platform. At the outer end of this platform were a water tank on a brick base, and the signalbox.

Once aboard the DMU ready for the return trip I picked a seat right behind the driver, because in those days they operated with the blinds up so that you got a marvellous view forward. To my delight, we were scarcely out of the station, when the driver slid back his door and invited me into the cab. I sat on the folding seat to his right and got an impromptu history lesson about the Calne branch, including the background to Black Dog halt and how it was built as the private station for the Marquess of Lansdowne, whose estate was nearby.

As he talked we sailed through Black Dog without stopping, and because it was approached through a cutting and round a bend I had no time to gather my camera to fire off a quick photograph. I did manage one of Stanley Bridge halt through the windscreen, as we passed non-stop through there, too. I recall there was a fairly steep drop down to a long girder bridge over the river just outside Chippenham, and then a sharp climb up towards the main line.

The arrangement at Chippenham was curious, at least to someone accustomed to branch lines which terminated in

Below:
Centrepiece of industry in Calne was Messrs C. & T. Harris's factory which produced bacon, sausages and meat pies. In this late 1950s view the incoming 'Harris pig special' is seen passing Black Dog with a '57XX' pannier tank in charge. At Calne the pigs would be unloaded and driven through the streets to the slaughterhouse. The factory site in Calne town centre is now a green. *D. Lovelock*

Above:
Harris's output was also shipped by rail, much of it in distinctively marked 'Siphon C' vans, three of which can be seen among the six vehicles attached to the rear of the 11.00am Calne-Chippenham auto-train. This is a Sunday morning working seen clearing the end of Calne station yard in the mid-1950s. *D. Lovelock*

Right:
The branch freight remained steam-worked almost until the end, which came in October 1964. In the latter months diesels were in evidence, however, and here BR 204bhp 0-6-0 No D2086 is seen heading a sizeable load away from Calne. *D. Lovelock*

A view showing part of Calne goods yard which shows just how busy it was, even in its declining years. It is worth noting that when freight services ended, Calne yard contained more than a dozen empty wagons, which was the maximum load for the available diesel shunter. Accordingly, the final freight working was actually a day later than intended. Clearly there was traffic on offer right up to the end.
D. Lovelock

neat bay platforms. The Calne branch turned westwards to join the down main at Chippenham East Junction. Branch trains then swung across to the right to terminate in a bay in Chippenham's centre island platform, for this station's unusual layout had the down main served by platform faces on either side.

This visit had merely confirmed that a longer excursion, specifically to Black Dog was warranted. The first attempt at this was an unmitigated disaster. The plan had been to travel by train to Calne, walk the mile or so from the station back to Black Dog, and catch a return train from there. Timetable-wise there were no problems. I still have my notebook from the period and in it I had noted that arrival in Calne would be at 15.08 and departure from Black Dog at either 16.05 or 16.53. There would, I reasoned, be ample time to walk there, have a look round, photograph the train on its down journey and catch the return working.

Between Black Dog and Calne the branch followed an almost continuous curve to the left. The effect of this was disorientating, to the extent that I turned the wrong way as I left the station approach at Calne. I had brought no lunch with me and was extremely hungry, so I went looking for something to eat. After all, I reasoned, in a town where the main industry was making pork pies, it ought to be possible to buy one for lunch. That was my second mistake. So, I came to leave Calne, hungry and walking up the A4 in the wrong direction. It was a long while before the awful truth dawned, and by then too late to correct my mistake. There was nothing for it but to abandon the plan and return direct from Calne.

Next time, and on all such subsequent trips I took the appropriate Ordnance Survey map. The exercise was successfully repeated a few weeks later and the same little notebook contains the hasty sketches I made at Black Dog and which formed the basis for the drawings which appeared in *GWR Country Stations — 2* and the late, lamented *Model Railway Constructor*. The only information missing from my notes is the exact date of the visit, but it appears to have been in the spring of 1965.

I left Reading on the 13.22 departure, headed by 'Western' diesel-hydraulic No D1005 *Western Venturer*. After a permanent way slack at Wantage Road and a 4min stop at Swindon from 14.06-14.10, we were into Chippenham at

Below:
Even after the dieselisation of passenger services some appropriate traffic was still attached to passenger trains. In the early 1960s it was not unusual to find a DMU such as this Swindon three-car cross-country set with a tail load. An ex-LMS parcels van trails the Chippenham-bound unit as it leaves Calne. *D. Lovelock*

Right:
A flashback to Great Western days and a scene which changed very little right up to the Beeching era. A Calne-bound goods is about to pass under the minor road bridge at Hazeland with a '20XX' pannier tank in charge around 1930. Most locomotives operated bunker-first towards Calne. The neat track and well maintained banks were a feature of most such branch lines, and on this one at least lasted right into the 1960s. Hazeland bridge is still there, and this stretch of the track bed is now the Marden Nature Trail, but views such as this would now be impossible through the dense woodland which has enveloped the area. *H. Gordon Tidey/Real Photographs (T6209)*

Below right:
A Derby-built suburban DMU creeps past permanent way staff at Black Dog halt on 1 November 1963 as work has commenced on lifting the siding. To the left of the train is the tiny wooden ground frame hut from which the points were operated. *D. Lovelock*

12

14.27. On this occasion the branch train was a Swindon-built three-car cross-country unit, Nos W50709, W59284 and W50652. These units had a small buffet counter in the centre coach but it was seldom if ever used, and certainly not on a five-mile branch line shuttle service! For the most part, the Calne branch passenger traffic could easily have been accommodated in a single car, but presumably there were insufficient of these to cover all the branches which needed them. Or, perhaps, there were certain occasions when three cars were needed.

This time it was destined to be a good day at Black Dog. On arrival at Calne I began by taking a good look around the goods yard. The freight service had been withdrawn and the yard was empty but the track and buildings were intact, even down to a grounded wooden van body of significant vintage adjoining the goods shed. Under Beeching, total closure had already been proposed and the green posters were up. They had been pasted over with a strip stating that closure was postponed pending a ministerial decision because objections to closure had been received. In those early post-Beeching months it was just possible to believe that lines might still be saved, although favourites like Tetbury and Staines West had already succumbed.

I love the English countryside and the walk out to Black Dog, past spring fields heavy with dew and under a misty sky, was pure magic. From the top of the hill, the halt could be glimpsed across the fields, almost hidden among the trees. The last few hundred yards were downhill to the entrance gates, just as I remembered from passing it in the car. The cream-painted wooden gates were broken and lay back in the undergrowth, and a stream gurgled through a culvert under the roadway. Beyond, an ivy-covered notice on bridge-rail posts read 'Carriage approach to the station. Foot passengers under the bridge', and a metal finger pointed to the footpath under the railway. The board was rotten and some of the cast metal letters missing or broken. I still have a letter 'E' which I picked up off the ground.

The gravel driveway which led up this side of the line was, however, far better than the near-impassable muddy footpath which foot passengers were directed to use. This was quite unlike any other halt I had seen on the Western Region. The platform, brick-faced and paved with slabs, was clearly of GWR origin but more akin to a station than a halt. The latter usually had timber-framed construction, or was simply a sleeper-faced mound surfaced with cinders. The building was unique, a little wooden shelter with an enclosed room at either end. By this time it was unstaffed and the rooms were locked. A Tilley oil lamp on an ornate metal post had been left burning all day.

The single goods siding had been lifted and sleepers were still stacked in the yard. Behind the waiting shelter, the footpath continued up over the hill, and from above the station there was a commanding view of trains approaching from the Chippenham direction. From here I photographed the DMU as it worked back to Calne, slowing over the A4 bridge in case there was anyone wanting to get on at the halt. There was not, so on went the power again and the unit slipped through the cutting out of sight. Regrettably, I had just got a new camera and was practising with black and white film. It would have made a nice picture in colour.

The Calne branch passenger service was withdrawn at the end of September 1965. I called there once more, at the beginning of that month, when returning from holiday. We were travelling by car and called briefly at Calne. A Derby three-car DMU was working the service. Just a single track remained at the terminus, the goods sidings and loop having been lifted. After that, I made one more visit to Black Dog, when the track was still *in situ*. Surprisingly, the halt had not been vandalised. At Stanley Bridge however, the cast letters

Below:
I have never thought of myself as much of a photographer, but this view of a Swindon three-car cross-country unit passing Black Dog on a misty afternoon is one of my favourite photographs. It was taken from a vantage point on the footpath immediately above the roof of the little station. I returned there one evening in May 1989 having visited Bowood House. Though the estate remains in the Lansdowne family and despite one Lord Lansdowne having been Chairman of the GWR there is surprisingly nothing whatever at the house to alert visitors to its unique connection with the little station with the strange name. As to Black Dog itself, a quarter century after Beeching there is depressingly little left to bring back the memories. *Author*

Right:
Calne station. By this time the freight service had been withdrawn and the goods yard was empty. The Harris loading dock can be glimpsed between the DMU and the goods shed. *Author*

Right:
There were no intermediate stations, or crossing loops, on the 5-mile branch to Calne through rolling Wiltshire countryside. Apart from Black Dog, the only other stopping place was Stanley Bridge halt, seen here in August 1964. It was a classic GWR halt, complete with 'Pagoda' shelter built by Joseph Ash & Co of Birmingham. Stanley Bridge served only a few farms and its main traffic had been in milk churns, for which a wooden shelter had originally been provided just beyond the waiting shelter, where the style of fencing changes. Note the well-kept trackwork only a year before closure. *Andrew Muckley*

STANLEY BRIDGE
HALT

Above:
Calne terminus with one of the Gloucester three-car cross-country units typical of the line laying over before the return to Chippenham. The main station facilities were in the wooden structure on the left, while the brick building (and beyond it a corrugated extension) were given over to handling parcels, produce, and for facilities for servicemen bound for RAF stations at Yatesbury and Compton Bassett. *Andrew Muckley*

had been removed from the nameboard and scattered over a field. We gathered them up for posterity and in 1988 they made a brief return 'home' for an exhibition in Calne to mark the 125th anniversary of the opening of the Calne branch.

This rural line, of just under five miles length, had given 102 years of valuable local service. With milk from Calne and Stanley Bridge, the Harris traffic and the extensive use by service personnel, not to mention a healthy volume of general merchandise traffic, it must have been one of the more commercially successful GWR country branches. Where there had been such success there must have been potential for development in other ways. Yet Beeching saw only the present balance sheet. Past performance and future prospects were not considered. Faced with Beeching's task one would have to admit that some closures were necessary — even inevitable — the Calne branch would not have been one of those choices.

It certainly *was* a bad day at Black Dog on 18 September 1965.

Below:
The view from the stop blocks at Calne on 9 September 1965, shortly before the branch closed and after all the goods yard track had been lifted. By this time the water tank had gone, the signalbox was out of use, and the station was unstaffed during the afternoons. *P. Strong*

CHAPTER 8

A Day Return to Chalford

Once the DMUs took over local services in the London area there was little opportunity to travel by steam unless one went further afield. Although we could still enjoy our local branches by diesel railcar it was inevitable that the attraction of steam would lead us westwards. By 1964 one of the last bastions of steam-worked push-pull operation on the WR was the Gloucester-Chalford service. This route had been the birthplace of GWR steam railmotor operations, and a number of small wayside halts had been built in the early years of the present century to serve lineside villages. Apart from the change to push-pull auto-trains things were much the same more than 50 years after the service had started. But Beeching had put the service up for closure so it was 'now or never' if we were to see the auto-trains in action.

At that time the main line to Gloucester was still busy. Cheltenham Spa was one of the Western's prestige destinations, and the roofboards bearing the legend 'Paddington Gloucester-Cheltenham Spa' still adorned a number of expresses each day. All that was also threatened with change. Beeching had listed the Swindon-Gloucester service to be 'modified'. That meant wayside stations would go and through services to London be eliminated or drastically curtailed. An 'improved' local service from Swindon-Gloucester meant, in reality, reduction of the route to almost branch line status, including even a single-track section between Swindon and Kemble.

No through trains stopped at Chalford, so our journey was planned with a series of changes. Staines to Reading and Reading to Swindon were uneventful. My notebook shows that we made an early start, leaving Reading at 08.47. 'Western' No D1064 *Western Regent* averaged 75mph

Below:
Surely one of the most evocative photographs of the Stroud Valley. '14XX' 0-4-2T No 1406 heads through the mist with a Gloucester-Chalford auto-working on 3 March 1953. *G. Heiron*

Above:
A scene very similar to the author's first sighting of the auto-train approaching Chalford through the cutting with its stone retaining walls. On 2 October 1964 '14XX' 0-4-2T No 1472 pants up the final yards to Chalford with an afternoon 'auto' from Gloucester. Note the signal with its sighting board, placed on the 'wrong' side of the line, a common GWR practice where sharp curves caused sighting problems. *F. A. Haynes*

Left:
In order to control a tiny gated level crossing on a very awkward site, the little signalbox at St Mary's Crossing cheated Beeching and MAS, and continues to stand sentinel over the run up to Sapperton. *Author*

between Didcot and Shrivenham before slowing for the 09.30 stop at Swindon. Here we changed to the Gloucester local, formed of just two or three coaches behind 'Hymek' No D7075. At this point my notebook is thin on detail. I was always a little haphazard in my note-taking, a habit which I have since regretted. Nor are any of my notes dated, although I have the ticket from this trip which shows the date to have been 21 March 1964. On reflection it seems odd that, with only two weeks to closure of the Tetbury branch we were prepared to pass Kemble by, but presumably the timetable did not allow us to do it all in one day. We did, however, arrange a change of trains at Kemble on the way home.

Locals to Gloucester stopped at Purton, Minety & Ashton Keynes and Oaksey halt before Kemble, and then over Sapperton to Chalford, Stroud, Stonehouse and Gloucester.

The intermediate halts beyond Chalford were only served by the auto-trains from Chalford, and for much of the route the line was paralleled closely by the main road, with an apparently reasonable bus service. Nevertheless, the Chalford 'autos' were quite well patronised and a good deal of surprise and alarm was expressed when the service came up for closure.

Purton station was a postwar rebuild, the main building on the up platform being a modern 'box' with a blue tiled finish. By contrast, Minety & Ashton Keynes which was some considerable distance from both the villages in its title, had original Brunel buildings in the red brick 'roadside' style. Goods facilities, signalbox and so on, had all gone by then.

Once over Sapperton, and out of the tunnel, the line emerged into the Golden Valley with the village of Chalford spread up the steep slopes on either side. It was still quite early when we arrived, with the sun just breaking through the early spring mist in this lovely wooded vale.

The train swept down into Chalford through a series of reverse curves, braking heavily to stop at the neat red brick station. I was unexcited by the standard GWR buildings, but an oil lantern with the station name on its glass caught my eye. We were the only passengers to alight. Two or three others boarded the train which then slipped off out of sight, although the throb of the 'Hymek's' engine could be heard echoing for some time. A look beyond the station revealed the line in a stone-walled cutting — perfect for echoes.

The railwayman on duty (was he the stationmaster, perhaps?) beckoned us across the lines by way of the board walk, to avoid the climb up over the footbridge. He clipped out tickets and we declined to buy any for the next stage of our journey, preferring to buy them on board the auto-train. There was a fair wait before the train was due, so we went for a walk and watched a couple of heavy freights come pounding up the bank, shoved in the rear by '41XX' 2-6-2Ts. Chalford was well placed, midway on the climb to Sapperton

Left:
There was lots of interest at Brimscombe station, which had original Brunellian buildings and was also the base for banking engines which assisted trains to Sapperton Summit. This is the view towards Chalford in August 1963 as an auto-train sets off up the bank. *Andrew Muckley*

Below:
Before the Beeching rationalisation Sapperton saw considerable freight activity. On 3 March 1953 a '28XX' 2-8-0 No 2819 toils up the bank above Brimscombe with a heavy train of stone. *G. Heiron*

It was not unusual to find '61XX' Class 2-6-2Ts as bankers at Brimscombe, but here No 6106 has charge of an up freight while 'country-cousin' No 5184 shoves hard in the rear as they pass Chalford. *P. H. Rigg*

Summit, and it was easy to appreciate the severity of the incline. 'Hall' No 6932 *Burwarton Hall* came pounding up, blasting steam from various leaky joints, with a train of loaded ballast hoppers from Tintern quarry, banked in the rear by 2-6-2T No 4104, the banker on duty from Brimscombe that day. A few minutes later the Prairie rolled back downhill light engine to await the next job. I could not have imagined then that I would stand train-watching on Sapperton 20 years later knowing that the only action I would see was a two-coach local DMU and a steam-hauled special!

Eventually the home signal just beyond the road overbridge crashed to clear again and our auto-train came into view, a grimy '14XX' 0-4-2T No 1451 at the head of two maroon auto-trailers. The locomotive was actually in fully lined out green livery under the dirt and the coaches looked smart and inviting in their lined maroon. The train came to a stand at the up platform to let passengers off, before heading further up the line towards the signalbox. Here, it stopped and reversed over a crossover to gain the down line and run back into the platform with the coaches leading.

The auto-trailers were of the so-called Hawksworth type, Nos W220-44, and entered service in 1951 and had contemporary bus-type seats. Entry was by way of an inward opening door, midway along the side and provided with retractable steps for use at certain halts with very low platforms. First disappointment to those of us brought up with DMUs was that the partition between the passenger saloon and driver's cab was of solid timber, so there was no forward view. We took up position on one of the longitudinal seats and contented ourselves with watching the shadow of our train on the lineside — most curious with the outline of the locomotive and smoke behind us.

The guard issued our tickets from a little wooden ticket rack with spring clip, just like the bus conductors used. Modernisation had yet to reach these trains, though just up the line the Tetbury and Cirencester branch guards had Setright ticket machines. Chalford to Stroud return was just 2s 2d (11p). The ticket was purple-coloured and carried the names of all the stations and halts along the route.

At Stroud we alighted. I seem to remember that we watched the next up auto-train go through, and waited to catch it on its return journey into Gloucester Central. Stroud had — indeed still has — attractive stone station buildings and a very fine matching goods shed. On the last few miles into Gloucester the Western and Midland lines ran side by side, leading to the occasional good-natured race between trains. This time the Midland was deserted.

Gloucester Central was a busy station, but from the enthusiast standpoint it was not very convenient. Long curved platforms made it very difficult to keep an eye on activities at both ends of the station, while Midland trains used an entirely separate station, Gloucester Eastgate, out of sight and connected to Central by a long covered footbridge.

During our stay we noted that three auto-trains operated the Chalford service, each one formed of two cars with the locomotive at the Chalford end and facing uphill for the climb to Sapperton. The coach combinations on that day were W230/4, W227/44 and W223/37, with '14XX' 0-4-2Ts Nos 1444/51/72. The only other notable feature of our visit to Gloucester was our first sighting of one of the new Swindon four-car InterCity DMUs.

We apparently made our way back to Kemble by local train and changed there to a Gloucester-Paddington express. The short stop at Kemble revealed the Tetbury branch being worked by AC Cars railbus No W79977, newly ex-works in the dark DMU green and with yellow warning rectangles. It was to be our last sighting of a Tetbury branch train as we could not face attending the 'last rites', when the service would be so artificially crowded.

Another 'Hymek', No D7052, took us back to Reading and stopped only at Swindon. Here, the platform stop of 8½min was by no means unusual. Western Region practice always seemed to require inordinately long station stops, nowhere more so than at Swindon. Nowadays, the tendency is for trains to stand at signals outside the station, or to crawl at snail's pace past the ugly wasteland that was once Swindon works.

Right:
Chalford station in 1951, with an unidentified '41XX' 2-6-2T banking a freight to Sapperton.
Ian Allan Library

CHAPTER

9 In Search of Stow-on-the-Wold

I think it was probably the name, as much as anything else, which prompted us to go looking for Stow-on-the-Wold. It was one of those lovely rural names like Bourton-on-the-Water and Ascott-under-Wychwood, which seemed almost certain to have an attractive station. Unfortunately the rail service had been withdrawn from this meandering cross-country route which linked Cheltenham to Kingham and Banbury.

Stow was the first station west of Kingham, and despite its tiny local population, Kingham had developed as a substantial junction station, the lines from Cheltenham and Banbury converging at its four platforms. For wartime traffic a direct link had been put in so that through Banbury-Cheltenham services could avoid going into Kingham, and this line formed the third side of a triangle, with Kingham station at its apex. Through the middle, and passing under this direct line, went the Oxford-Worcester main line.

Kingham was a curious case, for its importance railway-wise was out of all proportion to the village which it served. Yet this importance saw that the station survived after Beeching, despite the fact that its junction status had disappeared. Kingham, like nearby Charlbury and Moreton-in-Marsh, was seen as a railhead once branch services and local stations in the surrounding area had closed.

By road, however, Stow was nearer to Adlestrop than it was to Kingham, and in early February 1965 closure of the wayside stations on the Oxford Worcester & Wolverhampton Railway — the 'Old Worse and Worse' — was still a little way off. So here we had an opportunity to use the OW&W local service (thankfully the 'Cotswold line' title was still a long way in the future) and to walk to Stow. The return was going

Right:
The photographer is leaving Kingham aboard a Cheltenham-bound train headed by '41XX' class 2-6-2T No 4101, which will make its first stop at Stow-on-the-Wold. On the left a '94XX' 0-6-0PT No 9471 is marshalling a freight train for the same route. *Ian Allan Library*

Below:
The view westwards at Kingham with '94XX' 0-6-0PT No 8491 bringing a 'Toad' brake van along the up main. The lines towards Banbury and Cheltenham lead off to the right and left respectively, and the direct line joining the two crosses over the black girder bridge visible beyond the signalbox. In fact severe weather damage had already severed the line beyond Chipping Norton, so the word 'Banbury' has been removed from the large informative station nameboard. The two platforms to the right of the nameboard were used by Banbury and Cheltenham line trains. In the fork of the diverging lines, and hidden by the locomotive is the small red brick engine shed. *Ian Allan Library*

to be more of a problem because the frequency of trains did not allow for an easy return by the same route, and a lot of walking would have been involved.

An early start enabled us to take a short interlude at Handborough, so we left Oxford by Derby three-car DMU on a Saturday morning stopping service to Worcester Shrub Hill. Yarnton station had already closed, so Handborough was the first stop. A delightful little station with a typical Oxford, Worcester & Wolverhampton Railway timber station building, it managed to look picturesque even in winter time. In the summer it was set amid a colourful flower garden, while the station house with its equally attractive garden stood close by.

At Handborough we left the train, pausing only to photograph the up local DMU passing. Handborough station remained painted in the Great Western Railway's light and dark stone colours at this time and no one was then aware of the role that this little station would play later that same year, as the destination for the funeral train conveying Sir Winston Churchill to rest at nearby Bladon. Handborough station's nameboards helpfully displayed the additional information 'for Blenheim', being the nearest station to the great stately home once the Woodstock branch had closed. It was said that the doors in Handborough station were deliberately placed so that a straight red carpet could be laid through the building from the forecourt to the platform and right up to the carriage door. It must have been quite a knack to stop a train with the appropriate carriage door exactly opposite the carpet!

From Handborough we set out to walk to the next station, Combe halt, a timber-framed platform and shelter set on an exposed embankment. It proved a long walk because the general layout of roads in this area runs at right-angles to the railway. Thus, one has to walk away from the line, then turn parallel to it along a more substantial road, then turn back down the lanes to find the next station. The operation of a bus service would be difficult in these circumstances, involving a considerable mileage to visit all the places served by the railway. This was one reason why Beeching's closure bid was rejected. In fact, substitute bus services were not required to operate in perpetuity. There was a minimum period laid down, and comparatively few substitute services lasted much longer than that minimum.

The day of our trip was a typically grey, chilly February day and a strong wind was blowing. Combe halt was less than comfortable. A Tilley lamp had been left burning high on its wooden 'gibbet', and the wind rocked it wildly causing the flame to roar now and then. We feared it might come crashing down. In due course the 1.25pm off Oxford hove into view, the customary two coaches headed by a Standard '5MT' 4-6-0, its white smoke driven sideways by the wind. We were not sure whether it was a request stop so we gave ample warning with a hand signal, and felt slightly mean bringing the train to a stop just for the two of us.

When we stepped off at Adlestrop the weather was gloomy indeed, with a black and threatening sky. Just the weather for a long walk with little hope of any shelter if the heavens opened! Adlestrop station had only a corrugated pagoda

Above:
The photographer made his trip in May 1962, alighting at Bourton-on-the-Water, where No 4101 is seen waiting the arrival of a train from Cheltenham headed by sister Prairie tank locomotive No 5173. *Ian Allan Library*

shed as a waiting room on the down platform. The up side building matched Handborough in style, and had been smartly painted in Western Region brown and cream. But, where Handborough's setting was fairly open, Adlestrop was wooded, dark green and leafy, with lots of ivy and the ever-present babbling of the River Evenlode close by the down platform.

The day of our visit was not at all 'the afternoon of heat' of Edward Thomas's poem *Yes, I Remember Adlestrop*, but we had the advantage of being able to leave the train and go off to explore. Stow-on-the-Wold station, when we eventually reached it, was everything we might have expected. The road crossed over the line at one end of the station on a small bridge, the track bed beneath in a narrow cutting. As at Adlestrop, the surroundings were well wooded, the Great Western's fondness for planting evergreens along station approaches being equally evident here.

The whole station area was abandoned and derelict, the track having been lifted. In the empty goods yard a few piles of sleepers remained. The station itself had a neat Cotswold stone building on its single platform (there had been no passing loop here) and immediately beside it a small goods lock-up and an equally diminutive signalbox. The letters from the station nameboard lay scattered about. We hid them under a bush, thinking we might some day come back and collect them, but we never did. This little station had been designed to blend perfectly with its surroundings, yet the stone building was comparatively recent, dating only from 1936, when the original buildings here and at Bourton-on-the-Water were replaced by the GWR in matching style.

Everything else at both stations remained original — Bourton had a black timber goods shed — and both had Tilley oil lamps on concrete posts. Bourton station is nowadays a coal office, and the road now follows the line of the railway, while Stow is a (very) private dwelling tucked down a long driveway.

As we left Stow-on-the-Wold that murky afternoon the long walk up the hill into the village proved a graphic demonstration of how badly sited some of these rural stations had been. A long weary, uphill walk from the station might have been acceptable in Victorian days, but in the era of the bus and the motor car it was not. Once into the village itself, we merely had to find the bus stop (the Church Square, I think), and there was the bus waiting to take us back to Oxford. The very long single-deck AEC, in handsome City of Oxford maroon, black and 'duck egg blue' colours was quite a change for those of us used to London Transport's 'shorty' single-deck RFs.

Above right:
Returning towards Kingham, No 5173 passes the neat black timber goods shed as it heads away from Bourton-on-the-Water. The track layout, forming a loop through the goods shed was a common arrangement at rural GWR stations since it made for ease of shunting from either direction. Board crossings are provided for the signalman to reach the token exchange apparatus on the opposite side of the line. *Ian Allan Library*

Right:
Four years on the scene is a stark contrast. Along with the track and signalling, the waiting shelter and signalbox have gone, but everything else remains intact. The coal merchant has taken over use of the station building and continues in residence to this day — hardly surprising since at the time of its closure the station was little more than 25 years old. The surrounding site has, however, been completely redeveloped. *Andrew Muckley*

5173

COALS
OFFICE

CHAPTER 10 Travels to Tetbury

The trips down to Gloucestershire which had revealed the delights of Black Dog halt, also brought to our attention the Tetbury branch from Kemble. Our trips were always on Sundays so there were no trains on the branch and this seemed to add to its light railway style character. It was a single-track through rolling countryside, with a substantial terminus at Tetbury, much of it overgrown with long grass. On our second visit we managed to make a detour along the road which ran parallel with the branch, and turning off down a country lane, we came upon Culkerton. Here, surely, was the ultimate light railway station, looking for all the world like a set from a Will Hay film.

Set atop an embankment, the timber building stood on a framework of large timbers. The building itself was derelict, with broken glass, battened-over doors and bits leaning at all sorts of drunken angles. Grass grew in clumps on the platform, there was an open ground frame with just a couple of levers, and an equally derelict, though substantial, brick-built goods shed. This was a railway that had to be worth a visit, though for two schoolboys the first difficulty would be saving up the 23s 6d (£1.17½p) fare from Reading to Kemble.

It was one of the first long-distance trips we had taken. We found out that the service was operated by railbuses, and since we had never seen one, this was an added incentive. The date of our first trip seems to have been 25 January 1963, from the ticket which I still have. The day dawned cold, with a stiff hoar frost as we left home after an early breakfast, having cadged a lift to the station. It was the first of many visits involving a change at Swindon into a Gloucester-bound local headed by a 'Hymek' diesel.

We alighted at Kemble into the chill morning air. It was strangely empty and a mist hung over the station, crowding in from surrounding fields. It was, however, destined to be one of those beautiful winter days when, in the space of little

Below:
Lightweight '74XX' 0-6-0PT No 7418 poses with her crew at Cirencester Town in August 1950. *Real Photographs (K923)*

Right:
The signal has cleared for the Cirencester railbus to leave Kemble, having connected with the Gloucester-Swindon-Paddington train which is departing behind a 'Hymek' Type 3 diesel. The fine canopy over this platform has since been demolished. *Andrew Muckley*

more than an hour, the sun rolls the mist away, the air is cool and clear, and it is just great to be in the English countryside.

Kemble was a marvellous place, quite the best of the Great Western's country junction stations, and with the added bonus of two branch lines, that to Tetbury turning south from the down side, while the Cirencester branch came in on the up side. Indeed, even after all the post-Beeching rationalisation Kemble still retains some of the old charm. I was last there on one of the GW150 steam excursions with 'King' No 6000 *King George V*, when we had a long wait at Kemble for a late-running DMU to come off the single-track section. It was after dark and momentarily quite possible to deceive oneself that nothing had changed and the branches were still there. In the waiting room was a superb oak table. I am sure fellow travellers must have been a little curious when I got down on my knees and crawled under the table. Sure enough, there on the underside was the brand-mark, burned into the timber — 'GWR'.

To return to that morning in 1963. Both branch bays were empty when we arrived, save for a lone van parked at the stop block on the Tetbury side. At the end of the down platform was a large water tank of standard GWR type, but with a header tank mounted high above it on a lattice support. We had seen nothing like it elsewhere, and were later told that the tank also supplied nearby houses in the tiny village. Kemble's other association with water was the pump house situated close to the signalbox at the end of the down platform. Water from a local source was considered excellent for steam locomotive purposes and the pump house fed the supply by way of a pipe to Swindon. In the event of failure of the pump, a '57XX' class 0-6-0PT could be stabled on the adjacent siding and the pump would operate by steam from the locomotive through a connection rigged to the whistle valve.

The station buildings were stone-built in matching style and dated from the latter part of the 19th century, having replaced the original Tetbury Road station. The ends of the station canopies had the familiar glazed timber draught screens, and the theme of glazed screens was evident elsewhere, particularly along the Tetbury branch bay and around the triangular arrangement of buildings on the up platform. The latter were almost like cloisters, with a courtyard garden in the centre. Sadly, the rationalisation following closure of the branches included removal of the branch bay canopies and the associated screens.

An archway through the down side screening gave access to the Tetbury bay, a BR sign enamelled in brown and cream proclaiming 'Tetbury Train' with an arrow underneath. The bay was still fully signalled with semaphores and retained its run-round loop. Freight services still survived on both branches and during the day we were to see the little '16XX' 0-6-0PT make its transfer across from the Tetbury line to the Cirencester side complete with the appropriately branded 'Toad' brake van.

Kemble station had no goods facilities of its own, freight for the area being handled at Coates Goods, some distance further down the main line. There were, however, a couple of sidings on the Cirencester side of the station to enable the branch goods trains to be marshalled and any odd wagons held ready for the next working.

By the time that the railbus from Tetbury arrived, the sun had broken through. The old wooden home signal crashed to

'clear' and a couple of minutes later the little railbus came swinging round the bend, its flanges squealing against the curve. We knew of railbuses only by the Airfix model kit, but the Tetbury branch vehicle was altogether different, shorter and chunkier in appearance, spotlessly clean and with its polished waist strip glinting in the sun. The sliding door and bus-style seating were also rather novel as was the tiny 'cupboard-style' cab partitioned off in the left-hand corner of each end.

A wise enthusiast with a ticket collection would avoid booking through to Tetbury as this deprived one of the chance to buy a ticket on the railbus. This also contributed to the branch receipts, where a through ticket from, say, Reading would not have done. The bus-style Setright

Above:
The weed-grown Cirencester terminus, which once had a Brunel trainshed, was far too large for the little railbus, although passenger traffic was sometimes good, as many as 110 passengers having reportedly packed the railbus on market days. *Andrew Muckley*

machine carried by the conductor/guard issued a turquoise ticket showing the abbreviated name of the starting point. Flagstops at any of the halts were unlikely, so the railbus would be sure to reach its maximum 50mph at some point on the journey. The winding curves and long fixed wheelbase produced a curious motion which was felt most strongly at the ends of the car as a series of sharp sideways jolts. There was also the novelty of 'flagging' the train at a wayside halt. This was another practice taken from roadside 'request' stops, and came back to me vividly as I was able to practice it again in Western Canada some years later.

The ride, in a four-wheel vehicle over light branch line track with short rail sections, was best described as 'lively', and the little car had a good turn of speed. There were scarcely half a dozen passengers and we made no

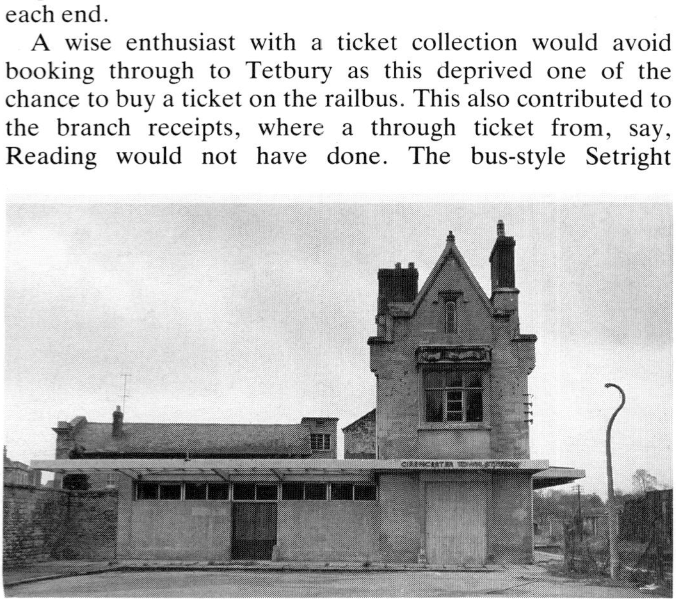

Left:
In its twilight years a half-hearted attempt was made to 'modernise' Cirencester Town by fitting a steel-framed canopy. In the Beeching era such expenditure was often a prelude to closure. Indeed, it came to be said that if the railway painted a station it meant it was going to be closed the next week. Cirencester Town is seen here in March 1966, two years after closure. It suffered the ultimate indignity of being converted into a bus station. *N. Caplan*

Above:
First stop out of Kemble on the Tetbury branch was Rodmarton, the only true GWR halt, complete with 'pagoda' shelter, on the branch. The railbus motors determinedly passed, apparently with a solitary lady passenger and the conductor/guard aboard. *Andrew Muckley*

Centre right:
Culkerton was the only intermediate station and boasted staff at the time of this early BR view. '45XX' 2-6-2T No 4550, still in GWR livery heads a typical mixed train of one coach and several wagons from Tetbury. Culkerton closed in 1956, but mixed trains operated and kept passengers waiting while they shunted goods wagons, almost to the end of steam on the branch. *Author's collection*

Right:
The remarkable outside-framed timber station at Culkerton was supported on stilts and was exceedingly derelict by the time it reopened to passengers in 1959, *R. J. Leigh*

intermediate stops. The arrival at Tetbury was all too soon. From there we walked the country road back to Culkerton, calling in along the way to visit Troublehouse halt.

A footpath led from the road opposite the Troublehouse Inn to this little wooden platform made of old crossing timbers, and little higher than the rails. It was said that the inn provided a beer crate to assist passengers boarding the railbus. There was a temporary permanent way slack here following some minor attention to the track.

Culkerton station was visible for some distance across the fields. A small girder bridge carried the railway over a lane

CULKERTON
KEMBLE

just on the Tetbury side of the station, and from here a gravel approach led up to the station and goods shed. Beyond it was a solitary house, presumably once having been the station-master's house and just about the only habitation in sight. We spent some time here sketching and photographing the station, though for the life of me I do not know why I took no colour photographs there. In due course we had a model of the branch, including Culkerton and one of the railbus halts, running round three sides of our bedroom.

Above left:
A lovely view of Culkerton, looking towards Tetbury, with the Kemble-bound railbus approaching. The platform was unusual in having paving slabs supported on a timber framework.
Andrew Muckley

Left:
By contrast to the station, Culkerton goods shed was a remarkably solid affair. By August 1964 tracklifting had reached this point. Beeching's wreckers lost little time and by 1965 it seemed that the bulldozers had more work than they could cope with. *Andrew Muckley*

Above:
With lights blazing the AC Cars diesel railbus bound for Tetbury approaches picturesquely named Church's Hill halt, which also served the village of Culkerton. The nameboard has been obliterated by the closure notice which was too big to fit and has been cut in half and pasted up sideways! Note the early concrete-sleepered track.
Andrew Muckley

Right:
The railbus is here seen posed at Troublehouse halt showing how the retractable steps worked at the low platforms. *Andrew Muckley*

Left:
Tetbury terminus in 1963 with the brick-built lamp room nearest the camera. The diagonal pattern roof slates were reused from an earlier station building, as was much of the timber joinery. Note the crude painted nameboard.
Andrew Muckley

Below:
A September 1964 view of Tetbury showing the goods shed and engine shed where the railbus had been shedded overnight. Contractors are recovering sleepers, the last vestiges of useable material to be removed before the site returns to nature. Beeching certainly was not the champion of the railways, but he might just have got the odd bouquet from those who seek to conserve nature or re-create the pre-Victorian era of transport in rural areas.
Andrew Muckley

Appendix 1

Passenger service withdrawals proposed in Beeching Report

A station name in *italics* indicates that the station had other passenger services and was not proposed for closure. Names in **bold** type indicate services which went through closure procedure but were reprieved and remain open.

Patney & Chirton-Holt Junction
Chippenham-Calne
Bath Green Park-Bournemouth West
Bristol Temple Meads-Bath Green Park
Taunton-Yeovil Pen Mill
Taunton-Minehead[1]
Tiverton Junction-Tiverton[2]
Taunton-Barnstaple Junction
Liskeard-**Looe**
Lostwithiel-Fowey
St Erth-**St Ives**
Kemble-Cirencester Town
Kemble-Tetbury
Gloucester Central-Ross-on-Wye-*Hereford*
Porth-Maerdy
Abercynon-Aberdare
Barry-Llantwit Major-*Bridgend*
Cardiff-**Coryton**
Caerphilly-Senghenydd
Bridgend-Maesteg-*Treherbert*
Aberystwyth-Lampeter-*Carmarthen*
Berkeley Road-Sharpness
Worcester Shrub Hill-Bromyard
Swan Village-Great Bridge
Old Hill-Dudley
Whitchurch-Welshpool
Ruabon-Dolgelley-*Barmouth*
Bala Junction-Bala
Llanymynech-Llanfyllin
West Drayton & Yiewsley-Staines West
Bristol Temple Meads-Portishead
Bristol Temple Meads-**Avonmouth Dock**
Yatton-Clevedon
Bristol Temple Meads-Patchway-Pilning
Bristol Temple Meads-Clifton Down-Pilning
Cardiff Clarence Road-*Cardiff General*

Below:
In the event, an act of God clinched the early closure of the Ruabon-Barmouth line when, in 1964, the line was washed out by flooding in the Dee Valley. The section between Llangollen and Berwyn has since reopened as a preserved railway. *J. Roberts*

Top:
Having dropped off only the mail and the single line token to the signalman, a Carmarthen-Aberystwyth train leaves Strata Florida station on 7 September 1963. This remote station boasted a corrugated iron building and one of the rare cast iron urinals. *Andrew Muckley*

Above:
From much the same viewpoint the station was photographed again in 1971 when it was destined to become a caravan site. *Andrew Muckley*

Right:
Sign of the times — but not for Beeching. *Andrew Muckley*

Below:

The sign which said it all. In the film *Titfield Thunderbolt* it was referred to as the 'death warrant'. The Severn Valley line between Kidderminster and Bridgnorth was luckier than most, finding a new lease of life in preservation. *Andrew Muckley*

BRITISH RAILWAYS BOARD

BRITISH RAILWAYS

LONDON MIDLAND REGION

PUBLIC NOTICE

With the consent of the Ministry of Transport, the British Railways Board hereby give notice that on and from Monday, 9th September, 1963, the passenger train service between SHREWSBURY AND BEWDLEY will be discontinued; BERRINGTON, CRESSAGE, BUILDWAS, IRONBRIDGE & BROSELEY, COALPORT, BRIDGNORTH, HAMPTON LOADE, HIGHLEY, ARLEY STATIONS, COUND, JACKFIELD, LINLEY, EARDINGTON, ALVELEY, AND NORTHWOOD HALTS will be closed to passengers.

The passenger train services between BEWDLEY AND KIDDERMINSTER and between BEWDLEY AND HARTLEBURY will be reduced.

Alternative passenger road services are already operated in the area by the following operators, from whom details of omnibus services may be obtained:–

Birmingham & Midland Motor Omnibus Company Ltd.
Wolverhampton Corporation Transport.
J. T. Whittle & Son, Highley.

Any further information required in respect of these arrangements can be obtained upon application to:–

Mr. E. R. WILLIAMS,
Divisional Manager,
London Midland Region,
43, Smallbrook, Ringway,
BIRMINGHAM, 5.
Telephone: Birmingham MIDland 5050 Ext. 303.

or Station Masters at:–

SHREWSBURY	Tel: Shrewsbury 3614
BERRINGTON	–
CRESSAGE	Tel: Cressage 312
BUILDWAS	Tel: Ironbridge 2185
IRONBRIDGE & BROSELEY	Tel: Ironbridge 2118
BRIDGNORTH	Tel: Bridgnorth 2208
HAMPTON LOADE	Tel: Quatt 221
HIGHLEY	Tel: Bridgnorth 2208
ARLEY	Tel: Bewdley 2242
BEWDLEY	Tel: Bewdley 2242
STOURPORT-ON-SEVERN	Tel: Stourport 2331
HARTLEBURY	Tel: Hartlebury 231
KIDDERMINSTER	Tel: Kidderminster 4004

The following ex-Southern Region lines came within the Western Region following the Beeching Report.

Okehampton-Tavistock-*Plymouth*[3]
Barnstaple Junction-Ilfracombe[4]
Okehampton-Padstow[3]
Okehampton-Bude[3]
Barnstaple Junction-Torrington[4]
Yeovil Junction-Yeovil Town
Chard Central-Chard Junction
Axminster-Lyme Regis
Seaton Junction-Seaton
Sidmouth Junction-Sidmouth[5]
Tipton St Johns-Exmouth
Exeter Central-**Exmouth**
Bere Alston-Callington[6]
Halwill-Torrington
Bodmin Road/Bodmin North-Wadebridge-Padstow[7]

Notes:

1 Minehead-Bishops Lydeard subsequently reopened under preservation as West Somerset Railway.
2 Tiverton Junction station closed 1986 when replaced by new Tiverton Parkway station.
3 Okehampton-Exeter services subsequently withdrawn and stations at Bow, North Tawton, Sampford Courtenay and Okehampton closed completely.
4 With these closures Barnstaple Junction became a terminus station.
5 Sidmouth Junction station was reopened as Feniton in 1971.
6 Bere Alston-Gunnislake section reprieved from closure as Calstock Viaduct provides a vital crossing of the River Tamar. Retention of this route also saved the ex-SR line from Plymouth to Bere Alston.
7 Bodmin Road was renamed Bodmin Parkway. Following the withdrawal of china clay workings between Bodmin Road and Wenford Bridge, the line between Bodmin General and Parkway was acquired for preservation.

Below:

Notgrove station, a scene of total dereliction in January 1966 at the height of the Beeching effect. *Andrew Muckley*

Appendix 2

Passenger services to be modified

A number of passenger services were listed for revision to eliminate uneconomic portions of services, reduce train mileage and enable underused stations to be closed. In general this meant the closure of local stations on main line routes, with semi-fast services calling only at major intermediate stations or 'railheads' as they were to become known.

Didcot-Swindon
Swindon-Bath Spa-Bristol
Bristol Temple Meads-Taunton
Chippenham-Trowbridge-Westbury
Reading General-Westbury
Bristol Temple Meads-Westbury-Weymouth Town
Taunton-Exeter St David's
Exeter St David's-Kingswear
Plymouth-Penzance
Par-Newquay
Cardiff General-Carmarthen
Carmarthen-Neyland/Milford Haven
Carmarthen-Fishguard Harbour
Swindon-Kemble-Gloucester Central
Cheltenham Spa-Cardiff General
Whitland-Pembroke Dock
Didcot-Oxford-Leamington Spa
Oxford-Worcester Shrub Hill
Stratford-on-Avon–Honeybourne
Stourbridge-Worcester-Hereford
Birmingham New Street-Worcester Shrub Hill
Worcester Shrub Hill-Gloucester Eastgate
Gloucester Eastgate-Bristol Temple Meads
Shrewbury-Welshpool-Aberystwyth
Salisbury-Exeter Central
Exeter Central-Barnstaple Junction

Appendix 3

Stations and Halts to be closed – individual stations not covered by closures listed in Appendix 1

Adlestrop
Aldermaston*
Ascott-under-Wychwood*
Ashley Hill
Ashperton
Athelney
Avoncliffe halt*
Aynho for Deddington
Bathampton
Bathford halt
Bedwyn*
Berkeley
Berkeley Road
Blaisdon halt
Bleadon & Uphill halt
Bletchington
Blockley
Box
Box Mill Lane
Bradford Peverell & Stratton halt
Bransford Road
Brent
Brent Knoll
Brimscombe
Bruton
Bugle*
Burlescombe
Cattistock halt
Chacewater
Chalford
Challow
Charlbury*
Cheltenham Spa Malvern Road
Chetnole halt*
Chipping Campden
Christian Malford halt
Churchdown
Churston
Coaley
Combe halt*
Corsham
Creech St Michael halt
Cullompton
Dartmouth
Dauntsey
Devonport Kings Road*
Dorchester West*
Doublebois
Dunball halt
Durston
Evershot
Exeter St Thomas*
Exminster
Fenny Compton
Fernhill Heath
Finstock halt*
Fladbury
Flax Bourton
Freshford*
Fritwell & Somerton
Grampound Road
Grange Court
Grimstone & Frampton
Gwinear Road
Handborough*
Hayle*
Hele & Bradnich
Heyford halt*
Holt Junction
Kidlington
Kingskerswell
King's Sutton*
Kintbury*
Lacock halt
Lavington
Limpley Stoke halt
Littleton & Badsey
Long Marston

Lostwithiel*
Luxulyan*
Lydney Town
Lyng halt
Maiden Newton*
Malvern Wells
Manningford halt
Marazion
Melksham
Menheniot*
Midgham*
Milcote halt
Minety & Ashton Keynes
Newnham
Norton halt
Oakle Street
Oaksey halt
Patney & Chirton
Pebworth halt
Pewsey*
Purton
Puxton & Worle
Quintrel Downs*
Roche*
St Columb Road*
Sampford Peverell
Savernake for Marlborough
Scorrier
Severn Bridge
Sharpness
Shipton for Burford*
Shrivenham
Silverton
Southam Road & Harbury
Sparkford
Starcross for Exmouth*
Staverton halt
Steventon
Stoke Edith
Stratton Park halt
Tackley halt*
Thornford Bridge halt
Tiverton Junction*†
Uffington
Wantage Road
Wellington (Som)
Weston under Penyard halt
Witham (Som)
Woodborough
Wootton Bassett
Wootton Rivers halt
Wyre halt
Yeovil Pen Mill*
Yetminster*

Wales

Aberaman
Abermule
Abertridwr
Beaver's Hill halt
Birchgrove halt
Bow Street
Briton Ferry
Caersws*
Caldicot halt*
Cardiff Clarence Road
Carno
Cemmaes Road
Clarbeston Road*
Clynderwen*
Cockett
Commins Coch halt
Coryton halt*
Ferryside*
Fishguard & Goodwick
Forden
Glandyfi
Gowerton North
Heath halt Low Level*
Jordanston
Kidwelly*
Kilgetty*
Lamphey*
Landore
Llanbrynmair
Llandre
Llanharan
Llansamlet North
Llantrisant
Magor
Manorbier*
Mathry Road
Montgomery
Pembrey & Burry Port*
Penally*
Pencoed
Penyrheol
Peterston
Pontdolgoch
Portskewett
Pyle
Rhiwbina*
St Clears
Sarnau
Saundersfoot*
Senghenydd
Skewen
Talerddig
Templeton
Undy halt
Welsh Hook halt
Whitchurch (Glam)*
Wolf's Castle halt
Ynyslas

Notes:
* Station either reprieved from closure or not actually proposed for closure subsequent to the Report.
† Station not closed until 1986 on completion of Tiverton Parkway Station.

Appendix 4

A reproduction of the table from *The Reshaping of British Railways* which featured a selection of sample assessments of financial savings which were expected to accrue from passenger service withdrawals. Interestingly, two WR services are considered and one of them is the Chippenham-Calne branch. Beeching did not feature the individual case for each service withdrawal in the Report and in most cases these were kept fairly secret and are subject to the '30-year rule' covering the release of public records. This air of secrecy at the time, coupled to the highly suspect figures which were published, fuelled speculation that few of these service withdrawals had any sound financial justification.

For instance, taking the Calne branch, the figure of £10,000 per annum (in 1963) for track and signalling maintenance on a five-mile branch with only two signals and two token instruments is very high. There is little or no explanation in the Report of what the figures covered and how they were arrived at. 'Terminal costs' is thought to cover staffing, yet Calne employed only one booking clerk at the end, and even if the train crew and permanent way staff were added, it is impossible to reach a wages bill of £8,900 per annum.

Local people, rail staff included, were adamant that the branch was still making a profit at the time of closure, and that freight traffic was still healthy when that was withdrawn. Apparently, the condition of a culvert near Stanley was quoted locally as a major factor in the closure decision. Yet this very minor culvert still stands a quarter-century afterwards. Perhaps the most ironic feature of the Calne branch table is the final column which concludes that closure would save £24,000 per annum — exactly the salary which the good Doctor was receiving as BR Chairman!

Below:
The financial logic behind specific closures

Region	Service	Traction	Type of service	Route miles	Train miles	Earnings £		Contributory gross revenue £		Expected loss in total gross revenue	Movement expenses		Net financial effect based on movement expenses	Terminal expenses	Total movement and terminal expenses	Net financial effect based on total movement and terminal expenses	Direct track and signalling expenses	Net financial effect based on total direct expenses
					'000	Total	Expected to be lost	Total	Expected to be lost	£	Total £	Per train mile s. d.	£	£	£	£	£	£
North-Eastern	Hull–York (via Beverley) ..	Diesel/ DMU	Rural/ commuter	42	259·7	90,400	64,790	37,680	4,900	69,690	84,400	6 6	+14,710	23,100	107,500	+37,810	43,300	+81,110
Scottish	Banff–Tillynaught	Steam	Rural	6	37·4	800	800	6,130	4,000	4,800	14,150	7 7	+ 9,350	950	15,100	+10,300	600	+10,900
North-Eastern	Bradford and Leeds to Skipton (via Ilkley)	DMU	Commuter/ rural/ holiday	26 and 28	404·2	59,700	59,000	24,370	5,930	64,930	73,900	3 8	+ 8,970	31,600	105,500	+40,570	48,400	+88,970
Eastern	Thetford–Swaffham	DMU	Rural	23	86·1	3,700	3,700	16,000	1,700	5,400	13,200	3 1	+ 7,800	3,900	17,100	+11,700	17,200	+28,900
Western	Chippenham–Calne	DMU	Rural	5	48·6	4,700	4,700	56,900	5,700	10,400	15,500	6 5	+ 5,100	8,900	24,400	+14,000	10,000	+24,000
Western	Yatton–Clevedon	DMU	Commuter	4	46·6	6,100	6,100	22,200	1,100	7,200	11,500	4 11	+ 4,300	5,600	17,100	+ 9,900	8,000	+17,900
Scottish	Elgin–Lossiemouth	Diesel	Rural	6	11·2	625	625	18,330	370	1,000	3,150	5 7	+ 2,150	1,500	4,650	+ 3,650	1,800	+ 5,450
North-Eastern	Sunderland–West Hartlepool ..	DMU	Inter-urban	18	38·4	8,500	5,300	3,370	170	5,470	7,400	3 10	+ 1,930	3,300	10,700	+ 5,230	..	+ 5,230
Scottish	Aberdeen–Ballater	Battery car/ DMU	Rural/ commuter/ holiday	43	161·0	14,370	14,370	21,980	7,000	21,370	19,850	2 6	− 1,520	9,250	29,100	+ 7,730	20,980	+28,710
Scottish	Gleneagles–Crieff–Comrie ..	Diesel rail bus	Holiday/ rural	15	65·1	1,900	1,900	12,280	9,000	10,900	7,500	2 4	− 3,400	3,480	10,980	+ 80	8,280	+ 8,360

The heavy vertical lines indicate the stage in the relationship between revenue and expenses at which withdrawal of the service becomes financially justified.
The direct track and signalling expenses shown above are the additional expenses incurred in providing track and signalling to passenger train standards for the services concerned.

Bottom:
The Calne branch goods leaves Calne through the misty meadows of Wiltshire. Note, to the right of the van, the field gate with unusual 'patent' tensioning post for the wire fence. Some of these, together with the GWR post and wire fencing, still exist along the earthworks of the branch. *D. Lovelock*

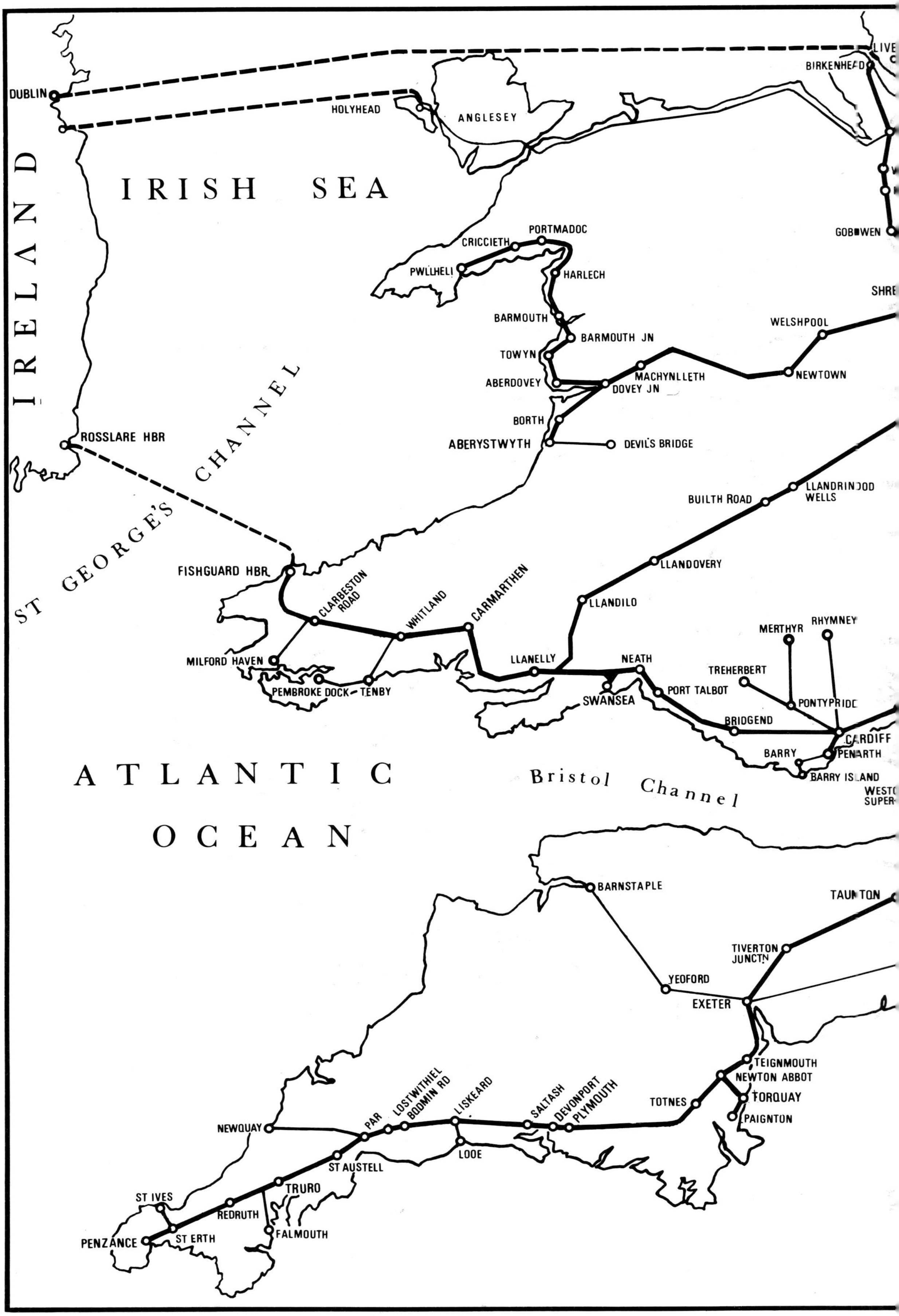
IRELAND
IRISH SEA
ST GEORGE'S CHANNEL
ATLANTIC OCEAN
Bristol Channel
ANGLESEY
DUBLIN
HOLYHEAD
BIRKENHEAD
GOBOWEN
PORTMADOC
CRICCIETH
PWLLHELI
HARLECH
BARMOUTH
BARMOUTH JN
TOWYN
ABERDOVEY
MACHYNLLETH
DOVEY JN
WELSHPOOL
NEWTOWN
BORTH
ABERYSTWYTH
DEVIL'S BRIDGE
ROSSLARE HBR
BUILTH ROAD
LLANDRINDOD WELLS
LLANDOVERY
LLANDILO
FISHGUARD HBR
CLARBESTON ROAD
WHITLAND
CARMARTHEN
MERTHYR
RHYMNEY
MILFORD HAVEN
LLANELLY
NEATH
TREHERBERT
PEMBROKE DOCK
TENBY
SWANSEA
PORT TALBOT
PONTYPRIDD
BRIDGEND
CARDIFF
BARRY
PENARTH
BARRY ISLAND
BARNSTAPLE
TIVERTON JUNCTN
YEOFORD
EXETER
TEIGNMOUTH
NEWTON ABBOT
TORQUAY
PAIGNTON
TOTNES
PLYMOUTH
DEVONPORT
SALTASH
LISKEARD
LOOE
BODMIN RD
LOSTWITHIEL
PAR
NEWQUAY
ST AUSTELL
TRURO
ST IVES
REDRUTH
FALMOUTH
ST ERTH
PENZANCE